20
59

THE FUTURE OF EDUCATION

MICAH SHIPPEE, PHD

Paperback ISBN: 979-8-9927231-0-6
eBook ISBN: 979-8-9927231-1-3

Cover design and typesetting by Enterline Design Services

For Laura, Bekah, Tripp, & Libby

Introduction: The Dawn of a New Educational Era

Education 2059

Orwell wrote his dystopian novel, *1984* in the year 1949, giving a glimpse 35 years into the future. In true Orwellian fashion, this book, *Education 2059* was begun in 2024. Statistically, I should be alive (at age 82) to defend this book or to at least stand against the fiery darts of my accusers.

The purpose of this book is to engage you, the reader, in a discussion about how we prepare our children for *their* future—a future dramatically different from our past experiences both educationally and professionally. A future that can be grounded and made more tangible by the recognition that innovation, in the context of education, is both an amplifier of our instructional practice (both good and bad) and a response to societal demands, including workplace preparedness.

In this book, I invite you to think ahead as we unpack the scalable tenets of good practice and strive to better understand how to best prepare ourselves to educate for the future. Throughout the narrative of this book I have chosen to use the inclusive language of "we" for you, the reader, and I to co-embrace the onus of creating a better future through our efforts in education. I believe it is, in fact, the responsibility of society to ensure our children are prepared for the world ahead.

Here we adopt the lens of a futurist, someone who has been exploring, analyzing, and discussing key trends, and possibilities, regarding the future. As we will see, historically, the future of work, specifically, has been a direct indicator of the future of education.

For those unfamiliar, some of the trends currently identified by futurists include:

- Automation and AI: Many predict that advanced AI and robotics will automate a significant portion of current jobs, leading to shifts in employment patterns and the need for reskilling.
- Remote and flexible work: The trend towards remote work, accelerated by the COVID-19 pandemic, is expected to continue, with more flexible work arrangements becoming common.
- Shorter work weeks: Some futurists advocate for three or four-day work weeks, arguing that they could improve work-life balance, productivity, and overall well-being. Several companies and even some countries have been experimenting with this concept.
- Universal basic income (UBI): As automation potentially displaces jobs, some futurists propose UBI as a solution to ensure a basic standard of living for all. This is related to the concept of universal wealth, though they are not exactly the same.
- Gig economy and freelancing: Many predict a continued rise in gig work and freelancing, with workers taking on multiple roles or projects rather than traditional full-time employment.
- Lifelong learning: With rapid technological changes, continuous education and skill updates are expected to become crucial for workers to remain relevant.

- Human-AI collaboration: Rather than full automation of everything, we should plan for increased collaboration between humans and AI, with AI augmenting human capabilities in various fields.

It is important to note that these are predictions and possibilities, not certainties. The actual future of work will likely be shaped by a complex interplay of technological, economic, social, and political factors. Yet undoubtedly these shifts will impact our education systems.

While I cannot pretend to know where exactly technological advancements will be in 2059, there are certain tenets to teaching and learning that we must diligently ensure our children have access to. I find myself, a tech early adopter, increasingly concerned about how technological advancements will impact our belief that we can and should create something unique, which is perhaps a fundamental truth of being human. Here another lens is that of a philosopher asking: What does that mean, "being human"? Philosophers, who emphasize free will and agency as fundamental to being human, often argue that these capacities are essential for:

- Moral responsibility and ethical behavior.
- Personal identity and self-conception.
- Human dignity and rights.
- Creativity and innovation.
- Social and legal systems based on personal accountability. [1]

It is important to think of these philosophical attributes as fundamentally human and that we must nurture them within our students to prepare for critical human processes, both now and in the future.

My Journey

As the author of *WanderlustEDU*, I am excited to reflect on how the evergreen principles explored in my first book can serve as a foundation for predicting and shaping the future of education. Wanderlust-EDU is about embracing the journey of innovation in education, and I believe it provides valuable insights for us as we investigate where education is headed.

In *WanderlustEDU*, I emphasize that innovativeness—the ability to continually create and adapt—is the true pedagogy of the future. This core philosophy underpins everything else in the book and would be crucial to consider in any work about the future of education.

WanderlustEDU explores how educators can navigate the ever-changing landscape of teaching methods and technologies by developing an innovative mindset and embracing change as a growth opportunity. The book presents practical models like the ARCS (Attention, Relevance, Confidence, Satisfaction) model of motivation and the SAMR model (Substitution, Augmentation, Modification, Redefinition) for technology integration to help educators understand and implement new approaches. It emphasizes creating an innovative school culture, explores emerging trends like personalized learning and global classrooms, and stresses the importance of adaptability, continuous innovation, and equipping educators with the mindset and tools to navigate future challenges and opportunities.

Addressing my biases

Some say the subtitle of every book is how to be more like me (the author). That being said, the onus is then on the author to be transparent about the critical lens they use to understand the field they are writing about. The following is my attempt to do just that.

I am the product of faith-based education. I am privileged to have received a humble education from dedicated, heart-led, educators of various backgrounds, who chose to give up their financial pursuits to educate children. I am forever and gratefully indebted to them and have pursued a life in education that I hope honors their sacrifice.

I believe it is the responsibility of education to help students on their journey to become lifelong learners with a fundamental focus on creating a safe, nurturing environment that deeply humanizes the learning process.

As a veteran educator, with decades of service to public education, my classroom experience informs my belief in what "works" in education. Further, my middle school teaching experience was guided by understanding my audience's perceptions and motivational needs. As a social studies teacher, I placed high value on the power of narrative, discourse, and research in the pursuit of ah-ha, eureka, lightbulb moments that, I believe, have resonated with my students and inspire them to be lifelong learners.

My work in higher education was grounded in instructional design principles, which point to teaching and learning that is designed and delivered with a learner-first approach. My work in this area can be simply thought of as prioritizing the following process to create instruction: analyze, design, develop, implement, and evaluate.

I have taught both undergraduate and graduate courses in Needs Assessment, Organizational Behavior, Principles of Adult Learning Theory, Planned Change, and Project Management, through instructional mediums ranging from online, hybrid, and in-person. From those experiences, I learned to embrace the importance of the technology being leveraged to teaching, appreciating the phrase, "the medium is the message." That being said, I also found my classroom teaching practice informed by decades of more physical teaching and coaching experiences in soccer, basketball, and downhill skiing. Here the power of tangible application of acquired skills became very real to me.

Finally, as a child of the 1980s I was highly attracted to the Space Race prompted by NASA, sci-fi and realized by Big Tech in the form of computing technology advancements. I was privileged to have access to Ataris, Commodores, Packard Bells, and the Turbo Graphx-16 (yes, I was that kid). Yet, I would have never thought of myself as a gamer, a programmer, or a true tech insider. Rather, I enjoyed these like I would have a board game. Something to be played with and then put away as I went outside in my sandlot-like life.

After 22 years of teaching in public education, I retired and joined Samsung as the Director of Education Solutions. While I believe it is important to be transparent about my role in the technology world, this book is my own opinion, not that of my employers, past and present. I have not written this book as a product pitch but I would be negligent to not share some context for my work, which has no doubt shaped my worldview.

In this book, I aim to explore upcoming educational trends and challenges, drawing on my diverse experiences and acknowledging my own biases and beliefs about what works in education.

Note: In this book we will discuss AI as a fundamental innovation shifting the future of education. AI is here to stay. The promise of streamlined workflows, a shorter workweek, and overall convenience in our daily lives has solidified AI's place in our world. In education, we strive to backwards-design curriculum and codify necessary skills from the future world our students will find themselves in. Preparing our students for their future is the critical focus of education and has always been our North Star.

Using This Book: Navigating the Future

Each chapter in Education 2059 concludes with an infographic and a structured summary designed to help you process, reflect on, and apply key concepts.

Chapter Infographics

The book's key themes and concepts are organized through a carefully structured visual framework that reflects both the progression of ideas and their interconnections. The organization follows several key principles:

Conceptual Flow

The layout progresses from foundational tools and frameworks through contextual understanding of change drivers, to deep explorations of key transformational areas, and finally to actionable recommendations and conclusions. This flow mirrors how innovation adoption occurs in educational settings - from understanding through planning to implementation.

Four-Point Structure

Each section contains exactly four key elements, creating a balanced presentation that helps readers focus on essential concepts. This deliberate constraint ensures that core ideas are highlighted while maintaining cognitive manageability.

The four-point structure also reflects the book's emphasis on comprehensive but focused approaches to educational transformation.

Unifying Themes

The bottom section identifies four overarching themes that weave throughout the entire framework: Technology as Enabler, Human-Centered Learning, Ethical Innovation, and Equitable Access. These themes serve as conceptual anchors, helping readers understand how various elements connect to create meaningful educational transformation.

This visual framework serves as both a roadmap and reference tool, helping readers navigate complex ideas while maintaining sight of their interconnections. The organization emphasizes that educational transformation requires understanding multiple perspectives and approaches, all working together toward the goal of meaningful learning in 2059 and beyond.

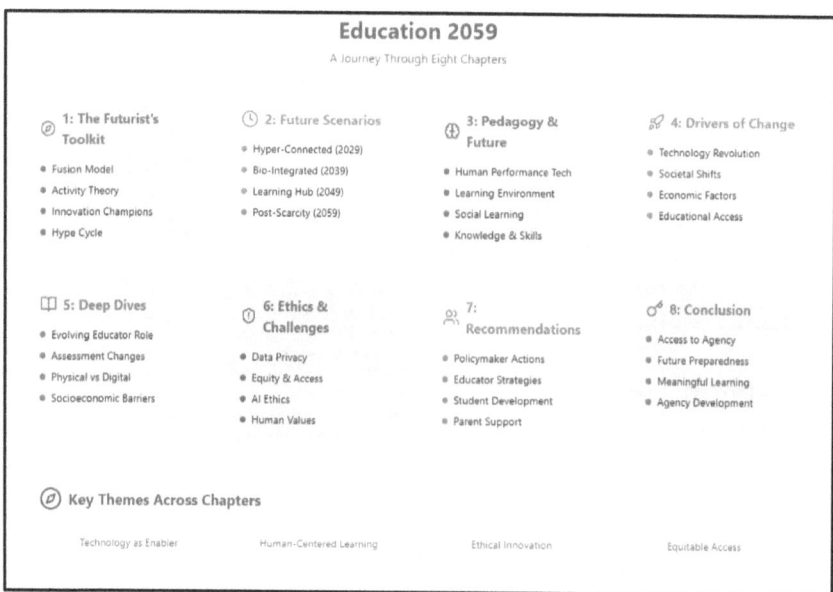

Chapter Structured Summary

Like early explorers who used celestial navigation to chart unknown waters, these summaries provide tools to help you navigate the future of education. Each summary contains three main elements: Horizon Scan, Waypoints, and Navigation Prompts.

HORIZON SCAN

Think of this section as your telescope—a broad view of the chapter's key themes and implications. The HORIZON SCAN provides a high-level overview that connects major concepts and points toward future implications. Just as sailors would scan the horizon to understand weather patterns and chart their course, this section helps you understand the broader context and direction of educational transformation.

WAYPOINTS

Like the fixed points navigators use to determine their position, WAYPOINTS highlight four key insights from the chapter. These insights serve as anchoring points for understanding and implementing change in your educational context. Each waypoint represents a crucial understanding that can guide your journey toward educational innovation.

NAVIGATION PROMPTS

These prompts serve as your compass, helping you chart a course forward. They are organized into three categories:

🎯 Implementation

- Practical questions about applying concepts in your educational setting

- Focus on immediate, actionable steps

💡 Innovation

- Questions about adapting and evolving practices
- Emphasis on creative problem-solving and new approaches

🌐 Impact

- Questions about broader implications and effects
- Consider long-term consequences and systemic change

At the end of each summary, you will find space for a FUTURE LOG where you can record:

- Observations from your current context
- Ideas you want to explore further
- Next steps for implementation

By engaging with these elements, you can move from understanding to action, creating meaningful change in your educational setting. Whether you're a policymaker, administrator, educator, or other stakeholder, these structured summaries provide a framework for processing information and planning implementation strategies.

Remember that like any journey into unknown territory, the path forward may not always be clear or linear. Use these summaries as navigation tools, but don't be afraid to chart your own course based on your unique context and needs.

THE FUTURIST'S TOOLKIT

OUR APPROACH TO PREDICTING EDUCATIONAL FUTURES

1. Explanation of futurist methodologies used in the book

Futurist writers employ certain methodologies to effectively communicate our ideas about potential future developments. In this book, I leverage these strategies to frame our understanding of the future of education:

Scenario planning: Presenting multiple possible future scenarios based on current trends and potential disruptors.

- Trend analysis: Identifying and exploring key trends that may shape the future.
- Forecasting: Making predictions about future events or developments based on data and analysis.
- Impact assessment: Evaluating the potential consequences of future developments on various aspects of society.
- Cross-disciplinary approach: Integrating insights from multiple fields to create a comprehensive view.
- Time horizons: Structuring predictions into short-term, medium-term, and long-term projections.

- Wildcards and black swans: Considering low-probability, high-impact events that could dramatically alter predictions.

2. Looking back to look forward

Education has always been a reflection of and response to the needs of society. To understand where education is heading, we must look to the past to see how it has evolved, what has proven effective, and what areas need work. The history of modern education is a story of constant change and innovation—a sometimes messy process of reform and reinvention.

In 1894, educator James Mackenzie wrote about the efforts of the Committee of Ten to reform education in the United States with the objective of preparing a productive future workforce. He noted, "We should recognize profoundly the need for reform in all parts of our school system." Mackenzie's report goes on to provide a fascinating historical lens that we can use to examine educational change.

The Committee of Ten recommended standardizing high school curricula and providing the same college preparatory program to all students, even though only a small percentage went on to higher education. They saw high school's main function as preparing students for "the duties of life." There was criticism that the American education system was producing too many "twenty-five-year-old babies," who were unprepared for adult responsibilities.[2]

The Committee of Ten's work led to the standardization of disciplines that created subject-area silos in secondary education—a legacy that still impacts cross-disciplinary instruction today. As we look to the future, we must consider how to break down these artificial barriers between subjects to better prepare students for the interdisciplinary nature of real world challenges.

History shows us that education is always a work in progress, constantly striving to prepare students for an uncertain future. By examining education's past transformations, we can better anticipate and guide its future evolution. The constant calls for reform and regular public scrutiny, while sometimes frustrating, is a necessary part of education's responsiveness to societal needs. Our role here is to learn from history's lessons as we iteratively improve our education system to empower students for the world they will inherit.[3,4]

In my experience, I have been diligently responsive to describe how recent innovations might be leveraged in various learning contexts, however with the fast pace of change we now live in it is time to start prescribing what we want the future to be for our learners. The time of waiting and seeing what comes next is over, it is time to embrace the conversation for what we want the future to be.... a dialogue desperately needed.

3. The importance of systems thinking in educational futurism

Techno-Economic Paradigms

As we look ahead to the future of education, it's crucial to understand the broader context of technological and societal change. Research by scholars like Carlota Perez has shown that approximately every 50 to 60 years, our world experiences a technological revolution that fundamentally reshapes not just our economy, but our entire social fabric.[5] These revolutions introduce new "techno-economic paradigms," essentially new, common sense models for how things should be done. However, the adoption of these new paradigms is not instantaneous or smooth. There is typically a period of mismatch between the new technologies and our existing social and institutional structures, leading to a turbulent phase of "creative destruction." This is where we find ourselves now, amid the fourth industrial revolution.

Perez's work also highlights a crucial aspect of technological revolutions that is particularly relevant to our current moment in education. She explains that primary technologies—like the large language models (LLMs) we are seeing today—always require a second wave of innovation. This secondary wave involves the development of applications and adjustments of organizational structures to fully leverage the new technology. To illustrate this, consider the historical example of electricity. The invention of the electric generator did not immediately revolutionize industry or daily life. It was only when electric motors were developed and production lines in factories were reorganized to take advantage of these new capabilities that electricity's true transformative power was realized.

Similarly, as we look to the future of education, we must anticipate not just the primary technological innovations, but also the wave of applications and structural changes that will follow. This pragmatic, tangible understanding can help us navigate the changes more effectively, reducing social costs and allowing us to harness the full potential of new technologies. As we imagine education for 2059, we must recognize that we are striving to not just adopt new tools, but to an entirely new paradigm in the future of work that emphasizes flexibility, networking, and decentralization. Our challenge is to responsively innovate our educational institutions in ways that align with this new paradigm, while still reflecting our core values and social choices. By doing so, we can create an education system that not only keeps pace with technological change but also helps shape a future that maximizes human potential and societal well-being.[6]

4. Innovation Champions and Technology Adoption in Educational Settings

A key component of this future-driven, change narrative is people. We can gain significant, strategic insight into how to help bring about positive change for a better future by leaning on research on how (and why) people adopt innovation.

The Role of Innovation Champions

In organizational change processes, innovation champions occupy a crucial position despite often experiencing imposter syndrome. This psychological phenomenon may actually drive their continuous pursuit of mastery. The essential competency for future success lies not in mastering specific tools or workflows, but in developing adaptability and change management capabilities.

Innovation champions are defined as individuals who play instrumental roles in navigating projects through institutional approval processes, demonstrating persistence, maintaining strong conviction in innovations, and engaging key stakeholders.[7] In the pursuit of organizational innovation, institutions can either cultivate internal champions or identify existing allies within these roles. Research indicates that organizational adoption of innovations increases significantly when key individuals (champions) actively support the initiative.[8]

The significance of this willingness to support change cannot be overstated. Organizations must actively counter the bystander effect[9], a psychological phenomenon where individuals assume others will take responsibility for addressing institutional challenges. This passive approach can significantly impede positive organizational change. Champions actively counter this effect through their commitment to institutional improvement.

Within organizational structures, champions function as sanctioned nonconformists, operating within institutional frameworks while maintaining the autonomy to explore alternative solutions. They often emerge as transformational leaders who cultivate support among organizational members.[10] Evidence demonstrates their substantial impact on both innovation adoption and sustainability. Organizations can enhance innovation adoption probability by identifying and empowering individuals who exhibit champion characteristics. Successful innovation champions typically possess specific attributes: They demonstrate credible expertise, exhibit strong planning capabilities, maintain effective collegial networks, and display sensitivity to peer needs while pursuing objective solutions. Additionally, these individuals must demonstrate tenacity, decisiveness, and the ability to inspire enthusiasm and optimism throughout the innovation process.[11]

Understanding Adoption Categories

The diffusion of innovations framework identifies five distinct adopter categories: Innovators, Early Adopters, Early Majority, Late Majority, and Laggards.

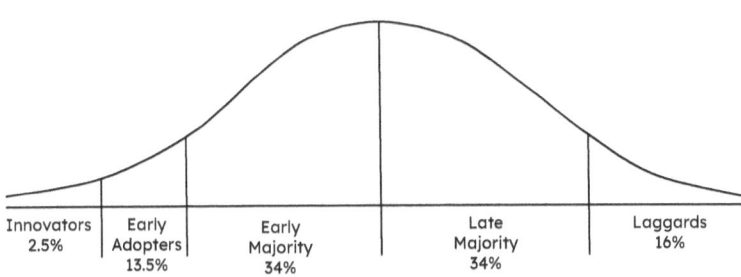

The Adoption Curve [12]

These categories can be visualized along a bell curve distribution based on innovativeness levels:

Innovator Characteristics

Innovators demonstrate strong interest in novel concepts from external systems, possess capacity for complex technical comprehension, and exhibit high tolerance for innovation-related uncertainty. Their approach to failure mirrors that of experienced gamers, viewing it as an learning opportunity.

Early Adopter Characteristics

Early Adopters maintain strong community standing and approach innovation with measured caution, preferring research-based implementation strategies. Change agents frequently engage this group to accelerate innovation adoption through local advocacy.

Early Majority Characteristics

The Early Majority adopt innovations just before the mean adoption time, serving as crucial links in successful diffusion. They implement established best practices based on Innovator and Early Adopter experiences, showing willingness to change while avoiding pioneer positions.

Late Majority Characteristics

Late Majority adopters follow the Early Majority, often motivated by economic or social pressures. They require demonstrated success before adoption and typically support organizational initiatives from a team-oriented perspective rather than personal enthusiasm.

Laggard Characteristics

Laggards typically maintain minimal community connections and evaluate innovations primarily through historical reference points. They require near-certainty of success before adoption and express significant concern regarding potential implementation failure.

Leadership Implications

When we dive deeper into the adoption dynamics it becomes clear that Early Adopters play a crucial role in innovation diffusion. While Innovators demonstrate initiative, Early Adopters validate innovations and reduce perceived adoption risks for subsequent adopters. This process highlights the importance of treating Early Adopters as co-leaders in change initiatives.

Additional key roles include opinion leaders, who exert informal influence, and change agents, who actively facilitate innovation decisions. Change agents systematically develop need recognition, establish information exchange, diagnose challenges, develop implementation strategies, and guide clients toward self-sufficiency.[13,14]

Decision-Making Framework

Organizational decision-making regarding innovation adoption typically follows three primary patterns[5]:

- Optional Innovation-Decisions: Individual-level choices independent of organizational consensus.
- Collective Innovation-Decisions: System-wide decisions reached through stakeholder consensus.
- Authority Innovation-Decisions: Determinations made by empowered organizational leaders.

Effective leadership strategies often involve allowing Innovators and Early Adopters flexibility in optional innovation-decisions, which can naturally catalyze collective innovation-decisions among Early and Late Majority adopters. Authority innovation-decisions may ultimately become necessary to ensure complete organizational adoption.

The understanding of these adoption patterns and decision-making frameworks remains crucial for educational leaders implementing technological innovations. Individual adoption patterns may vary across different technologies, reflecting the complex nature of innovation diffusion in educational settings.

The Fusion Model

Planning to introduce an innovation is a change management process that in and of itself requires strategic thinking.

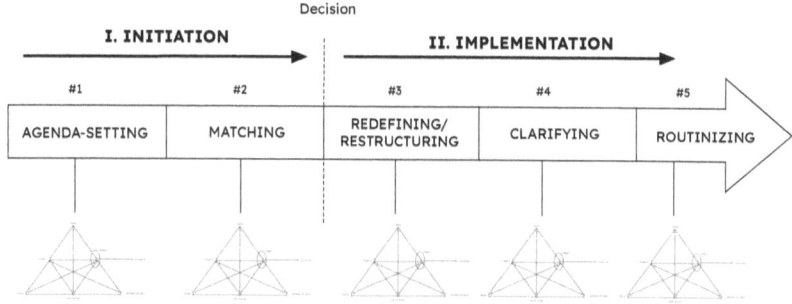

The Fusion Model [15]

Change is inevitable. Yet change is often painful. The Fusion Model explains how we can look at where we are and what needs to occur for us to successfully adopt innovation which will promote positive change.[16]

The Fusion Model was developed to analyze innovation adoption and provide a comprehensive framework for evaluating new tools beyond their immediate usability. In education, this model encourages us to consider how an innovation fits within the broader context of a school's culture, existing systems, and long-term goals. A model like this prompts reflection on how the innovation will disrupt current practices (both positively and negatively), require restructuring of workflows, or necessitate professional development. By applying the Fusion Model, we can better anticipate the full scope of adopting a new technology, including potential challenges and organizational changes ensuring that decisions are not based solely on short-term impressions, but on a thorough understanding of the adoption process and its implications for the entire educational ecosystem.

Why Fusion?

Admittedly, I am a bit of a history nerd, the name fusion celebrates two Cold War-era theories developed on both "sides" and are fused together. Additionally fusion, like adoption, represents something new emerging through the alignment of concepts, objectives, and goals. The Fusion Model was built on two research fields: the diffusion of innovation and activity theory. Diffusion of innovation research examines how practitioners approach new initiatives, how they attempt to adopt new methodologies with new technologies into their local contexts, and what learning challenges they are facing and why.[17] The Activity Theory (AT)[18] system illustrates how the multiple variables in any activity are involved in the generation of a specified outcome. The AT system provides a critical lens to investigate the activity of innovation adoption within an organization. Both were developed independently of each other and have been brought together (fused) to support the disruption caused by the pursuit of meaningful change.

What is Innovation?

Change in education often comes through catalysts like new research-based strategies or new technological achievements in the workforce that we believe our students must be prepared for. These changes are, in fact, innovations. An innovation is an idea, practice, or object that is perceived as new by an individual or organization. We often think of technology being the very definition of innovation, yet technology is simply a *type* of innovation. The process of adopting innovation often begins when we engage in opportunistic surveillance by scanning the environment for new ideas that might benefit our educational settings (the organizations impacting learning). Sometimes knowledge of an innovation, rather than the recognition of a problem, launches the innovation process. It is in this case that we remember to pause and reflect on our "why"... that is, our goal.

Why Adoption (Rather Than Integration)?

As we will see, to fully leverage an innovation, it must be adopted into our processes at the individual user level. In this book, I intentionally use "adopt" rather than "integrate" to understand the full impact an innovation can have on an organization. Adoption and integration are very different terms, *adoption* implies the power to choose, whereas *integration* can infer a forced change.

5 stages of Organizational Adoption of Innovation[19]

Organizational adoption can be viewed as a left-to-right process from initiation to implementation. Each stage describes the steps taken by an organization adopting an innovation. Once the criteria for a stage is satisfied, the organization advances to the next stage.

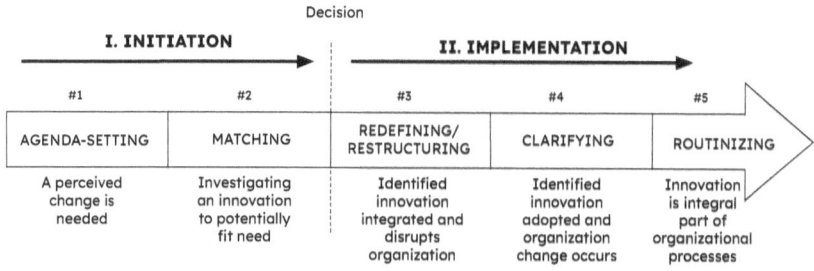

The Fusion Model: Stages Defined
(based on Rogers Diffusion of Innovation in Organizations[20])

When we begin to choose change, we are in an Initiation phase where we set our Agenda and begin to Match our findings with our purpose. The Agenda is our way of describing that there is a need for change. Next, we explore innovations that we believe Match our needs. Remember, an innovation can include a tool or strategy. Once we decide what innovation we think it is best to proceed we enter the Implementation phase.

Implementing an innovation begins with Redefining/Restructuring the innovation to fit our structure. We must prepare ourselves for our school hierarchy to change as well, innovation is a disruptor that can lead to change at many levels including in our routines, processes, and practices. As we progress through the Implementation phase of an innovation we next Clarify how the innovation is perceived, or defined, by our school culture. When we no longer think of an innovation as independent of our school identity, we have Routinized the innovation and it is simply just part of what we do.

At each stage of the process, failure to initiate or implement an innovation was a real possibility, as with any emergent technology comes unforeseen problems, and the strategies used to address them can greatly inform future decisions. For more about the organizational adoption process see the Appendix.

Activity Theory: Remembering our Why

The critical lens employed in the Fusion Model emphasizes a focus on our organizational activities as they relate to our goal. When one attribute is changed, the rest must adjust to ensure the goal is still achievable.

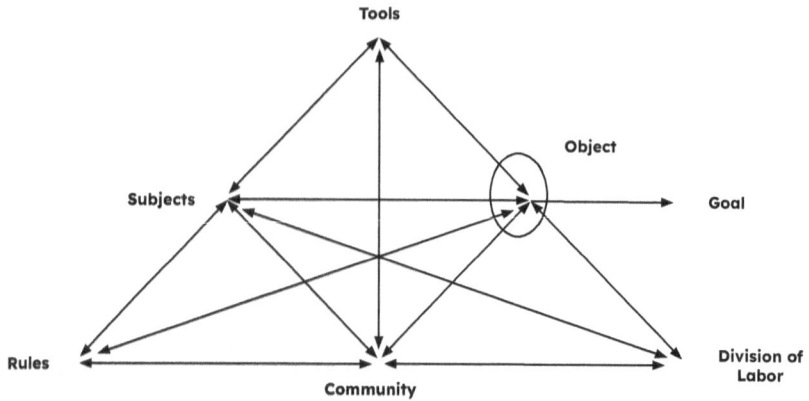

The Activity System

The Activity Theory (AT) system illustrates how the multiple variables in any activity are involved in the generation of a specified outcome. The AT system provides a critical lens to investigate the activity of innovation adoption within an organization.

The final step, the interaction of the components, represented by arrows, of the system are examined to reveal any contradictions or inconsistencies that will impact the achievability of an organizations' goals.

Activity Theory serves as a critical lens to understanding your organizational context.

AT System Components Definitions and Practical Applications

	Defined	In Practice
Subject	The individuals and work groups that would be formed in the organization to fulfill goals through the activity of instructional design and development. Individual actions include conducting a needs assessment, performing task analysis, and designing instructional interactions. In an instructional design context, this may be a single designer or a team consisting of designers, a manager, subject matter expert, and media producers.	• How do you understand your organizational culture? (Who fits where from your perspective?) • FOCUS: Who are the innovators and early adopters who can help? What are their roles?
Object	Good instructional practice is dynamic, changing to meet the needs of our learners. In an instructional design context, this may be a curriculum design, hypertext program, workshop, or a videotape that is produced.	• What types of interventions might work to help organizational adoption?

Tools	The instructional tools we use should amplify good practice with the outcome of knowledge retention and/or skill acquisition. In an instructional design context this may consist of the software production tools, project management system, or any other kind of tool that instructional designers use to transform the object.	• What technology based solutions are available?
Community	Consists of the people of the entire educational setting that share a set of social meanings.	• How would you describe your organizational community? Locally? Regionally?
Rules	The inherent guide of the actions and activities acceptable by the community, with signs, symbols, tools, models, and methods that the community uses to mediate the process.	• What means (financial and otherwise) are acceptably accessible for completion of the project? • What methods are used to mediate the process? What departments are used to support adoption?
Division of Labor	Prescribed task specializations (designers, developers, producers) by individual members of groups within the community or organization.	• What does the organizational hierarchy look like? • What tasks need to be achieved?
Goal	The form of instruction that is developed and implemented.	• What will be produced?

This critical lens is employed as a stage-snapshot of where we are, where we have been, and where we want to be. The entire process of change is challenging and can be complicated, the Fusion Model is designed to help us plan for change, a fundamental need, as we approach the future of education.

Employing the Fusion Model: A Case Study

The Fusion Model fuses the magnifying glass (critical lens) of the Activity System together with our understanding of the stages of the innovation process in an organization.[21] The stages help us to describe what is happening in our schools as we attempt to embrace change throughout the iterative nature of our study.

Perhaps the easiest way to understand the Fusion Model in action is through an example or case study.

School 1234: Leveraging AI - A Case Study

Imagine a school, we can call it School 1234, located anywhere in the world. The administration believes the school needs to prepare students for an AI-driven future while also leveraging AI to enhance learning (a tool shift). The school wants this to align with their established future-ready goals. Here's how School 1234 progresses through the innovation process stages:

Agenda-Setting

To achieve this goal, the administration begins to pursue increased AI integration across the curriculum and school operations. They want the changes to match their desire to prepare students for the future while enhancing current learning experiences.

Matching

After investigating best practices, the administration believes a multi-pronged AI approach will be most effective:

- Integrating AI literacy into the curriculum.
- Using AI tools to personalize learning.
- Leveraging AI for administrative tasks.

Teachers identified as innovators and early adopters are brought into the dialogue to address initial concerns (Teacher Advisory Team and AI beta-user Educator Committee). The need for extensive professional development is identified due to the impending disruption.

Redefining/Restructuring

The rollout of AI tools and curriculum changes begins:

- AI literacy is integrated across subjects.
- Personalized learning platforms using AI are introduced.
- Administrative AI tools for tasks like scheduling are implemented.

Professional development is provided to help with the pedagogical shift. Innovators, early adopters, and early majority attend initial training sessions.

Clarifying

The administration announces mandatory AI training for all staff. They demonstrate commitment by allocating a significant budget to AI tools and training. An AI ethics committee is formed to address concerns. Ongoing professional development supports late adopters and laggards in shifting their practices.

Routinizing

Eventually, AI literacy, personalized AI-driven learning, and AI-assisted administration became standard practice at School 1234. The school no longer identifies as an "AI school"—it's simply how they operate. The norm for their school culture has shifted.

Has School 1234 finished innovating? No! To sustain a culture of innovativeness around AI, they must:

- Continuously evaluate new AI developments for potential integration.
- Maintain an active AI ethics committee to address ongoing concerns.
- Provide regular opportunities for all stakeholders (teachers, students, parents, community) to give input on AI use.
- Offer ongoing professional development as AI capabilities evolve.
- Regularly assess the impact of AI on learning outcomes and make adjustments.

School 1234 must continue cycling through the implementation phase as they pursue their goal of ethically leveraging AI to prepare students for the future while enhancing current education. The key is viewing AI integration as an ongoing journey of ethical innovation, not a final destination.

This seems to illustrate a clear picture, but when we add the Activity System to the equation, we start to see a much richer explanation of what is going on and what shifts occur. If we were to zoom into each of these stages to see what was going on in our school, we would need a way to understand the chaos. The Activity System gives us an excellent way to dissect the interactivity involved with all that is going on around our school's choice to innovate.

The Fusion Model: School 1234 Leveraging AI

	I. INITIATION		Decision	II. IMPLEMENTATION		
Activity System	**#1 Agenda-Setting**	**#2 Matching**	**#3 Redefining/ Restructuring**	**#4 Clarifying**	**#5 Routinizing**
Goal	Prepare for an AI-driven future	Prepare for an AI-driven future	Prepare for an AI-driven future	Prepare for an AI-driven future	Prepare for an AI-driven future
Community	Future Ready	Future Ready	Future Ready	Future Ready	Future Ready
	K-12	K-12	K-12	K-12	K-12
Objects		Teacher Advisory Team	AI Professional Development	AI Professional Development	AI Professional Development
Rules	Pre-defined curriculum prescribes instruction	Pre-defined curriculum prescribes instruction	Pre-defined curriculum prescribes initial instructional planning	Pre-defined curriculum prescribes initial instructional planning	Pre-defined curriculum prescribes initial instructional planning
	Digital tools for manual workflows	Digital tools for manual workflows	Digital tools available	Digital tools available	Digital tools available
		AI used for automation of some workflows	AI used for automation of most workflows	AI used for automation of most workflows	AI used for automation of most workflows
		AI used to dynamically inform curriculum (describing student progress)	AI used to dynamically inform curriculum (describing student progress)	AI used to dynamically inform curriculum (describing student progress)	AI used to dynamically inform curriculum (describing student progress)
Division of Labor	Curriculum committees	Curriculum committees	Curriculum committees	Curriculum committees	Curriculum committees
	Admin Team	Admin Team	Admin Team	Admin Team	Admin Team
		AI beta-user Educator Committee	AI beta-user Educator Committee	AI ethics committee is formed	AI ethics committee
Tools	Traditional Curriculum	Traditional Curriculum	Traditional Curriculum	An "AI school"	Curriculum
	Traditional School Operations	Traditional School Operations	Traditional School Operations	AI literacy is integrated across all subjects	School Operations
		Integrating AI literacy into curriculum	AI literacy is integrated across all subjects	Personalized learning platforms using AI adopted	
		Use AI tools to personalize learning	Personalized learning platforms using AI are introduced	Administrative AI tools for tasks like scheduling are implemented	
		Leveraging AI for admin tasks	Administrative AI tools for tasks like scheduling are implemented		
Subjects	Teachers	Teachers	Teachers	Teachers	Teachers
	Students	Students	Students	Students	Students

ROUTINIZING IS WHEN WE DROP THE "AI"!

With the Fusion Model, we take the five stages of the innovation process in an organization, and use them to frame our understanding of the process, we then explore the impact on our school activity. When one adjustment to an Activity System component occurs, it affects the relationship each component has with the other. For example, in the chart when AI (Tools) are introduced, beta-users and the ethics committee (which will need to ensure authenticity of learning metrics and assessment) become critical to achieving the school's goal: Preparing students for an AI-driven future, the AI beta-users Educator Committee and AI Ethics Committee represent a shift in the Division of Labor.

We should not expect our school activity to stay the same as we engage in the innovation process. There is just too much happening for us to remain the same. Innovativeness is embracing this constant change. The Fusion Model gives us a way to think about all that is happening.

Our schools are more dynamic than many of us are comfortable to admit. We cannot assume decision making to be happening in isolation, rather our decisions should be dynamically-refined, thus triggering consequent changes in school activity and the way we function. But we must always remember, the more things change, the more they stay the same. When it comes to the process of adopting innovation, this rings so true.

To develop a sustainable level of innovativeness, we must understand that we will be constantly redefined and reinvented; we must promote access and employment of meaningful technology; and we must find, embrace, and elevate our champions.

It is difficult to predict the future of learning, however, from the perspective of Moore's Law, future technological innovations will only become more powerful and conveniently part of our daily lives. It is important to understand how we can leverage new innovations to effectively and efficiently achieve instructional communication goals that transcend time and space. Without this understanding, trending technology will be limited to entertainment and anecdotal uses.

For our school culture to achieve innovativeness we must understand that it is a choice. It is up to each of us to choose whether or not to adopt (meaningfully) emergent technology. We sometimes find ourselves the recipients of shared decision-making, where decisions are made, and then shared with us. This is not a very motivating situation, but the core of innovativeness is choice, no matter how, who, or why decisions are made. The choice is to adopt or reject an innovation that is made by us as individuals, independent of the decisions by other members of our school. We can be influenced by the norms of the system (those around us) and by communication through interpersonal networks.

Innovations change our organizational structure. The unsettling nature of the diffusion of innovation process is supported by other researchers who understand innovation as organizational change.[22] Our schools can best approach innovativeness as a disruptive change. The main purpose of the Fusion Model is to explore the change process by which we adopted and engaged with innovations in our schools.

With a plan in place to approach change, we must then look to the future to understand what is changing, the actual innovations of the future employing logic, reason, our understanding of human nature, and predictive frameworks like the Hype Cycle.

Hype Cycle

The rate of change in technology, innovation, and policy is accelerating at Moore's Law-like speeds. In education we must seek tools to help us understand how to best invest our time and efforts in investigating these changes. Readers of this book may identify as:

- Credible experts in edtech
- Curriculum coordinators
- Education policy makers
- School administrators
- Educators
- Pre-service teachers
- Students
- Parents
- School planners and designers
- Concerned citizens

…and many others, each with a potential track record of leading, supporting, and applying their learnings in education. Here, and throughout this book, I use "we" to speak to the common mission all of the above have to keep our learning spaces relevant in preparing for our shared future. We use our research and networking skills to effectively and efficiently recommend changes to keep in education understanding the challenge we face to stay on top of emergent technologies as they relate to education.

One important method of exploring emergent technology is to focus on what is, and what is going to be, relevant for our students. I have found help through exploring the Gartner Hype Cycle for Emerging Technologies, which is released every year to describe the business potential of new technology.

While assessing current needs is crucial, we must also anticipate future skills and competencies essential for our students. The
rapid pace of technological change means today's "needs" may soon
become obsolete. I recommend combining a needs assessment of
current gaps with a futurist perspective on future requirements. This
dual approach enables us to make strategic technology decisions,
creating adaptive learning ecosystems that remain relevant in a
changing world. The future of education is about cultivating skills
for lifelong learning and innovation. Technology should enable this
broader vision, not be an end itself. By balancing current needs with
future projections, we can better prepare students for upcoming
challenges and opportunities.

Research is critical and should be a mix of use cases from our
trusted network, however, I would emphasize the value of pushing
ourselves beyond the dangers of groupthink and extend our research
to leverage tools like the Gartner Hype Cycle into this process. The
Hype Cycle provides a critical framework for understanding the
maturity and adoption of emerging technologies. For example, in
WanderlustEDU, I wrote about a Hype Cycle report that pointed us
to a future where AI was everywhere.

The Hype Cycle offers a long-term perspective that complements
immediate needs assessments. It helps us distinguish between temporary hype and genuinely transformative technologies. This insight
is crucial for building a sustainable, future-focused educational
technology strategy that balances innovation with practicality.

In education it is our job to prepare students for this rapidly
accelerating world. We have to be prepared to rapidly adapt to the
relevant workforce changes.

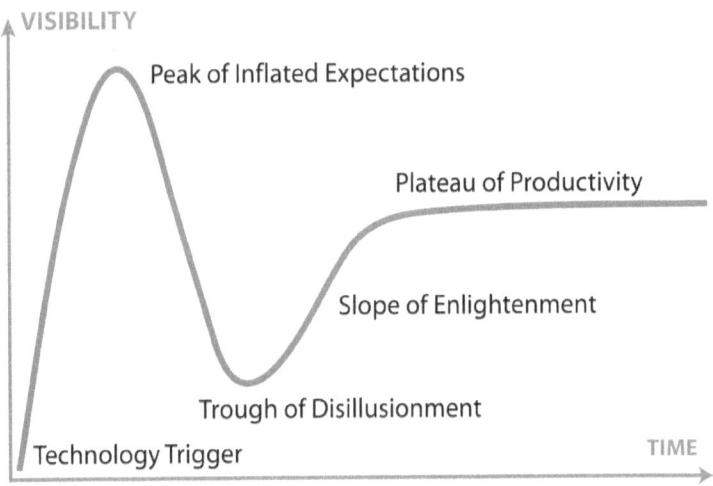

The Hype Cycle

The x-axis of the cycle describes time-related concepts which need further elaboration:

1. Technology Trigger: A potential technology breakthrough kicks things off. Early proof-of-concept stories and media interest trigger significant publicity. Often no usable products exist and commercial viability is unproven.

2. Peak of Inflated Expectations: Early publicity produces a number of success stories—often accompanied by scores of failures.

3. Trough of Disillusionment: Interest wanes as experiments and implementations fail to deliver. Producers of the technology shake out or fail. Investment continues only if the surviving providers improve their products to the satisfaction of early adopters.

4. Slope of Enlightenment: More instances of how the

technology can benefit the enterprise start to crystallize and become more widely understood. Second- and third-generation products appear from technology providers. More enterprises fund pilots; conservative companies remain cautious.

5. Plateau of Productivity: Mainstream adoption starts to take off.

In short, this is how the financial institutions and tech world thinks about the future of work and how innovations impact societal change.

As we envision the future of education, let's pause for a moment to consider how innovation has already shaped us. Just as early explorers used the stars to navigate uncharted waters, we too have tools to help us chart a course through the ever-changing landscape of innovation.

The Hype Cycle is a map that illustrates the progression of an emerging technology from its trigger point through inflated expectations, disillusionment, enlightenment, and eventually to productivity.

The latest Hype Cycle predictions for education paint an intriguing picture of the road ahead. In 2017, AI, extended reality, and adaptive learning platforms were positioned as transformative technologies climbing toward the "Peak of Inflated Expectations."[23] Since I discussed this prediction in *WanderlustEDU* (2019). AI has dramatically impacted the world with no signs of slowing down. As innovation champions, we must approach these with both enthusiasm and a critical eye.

Consider AI-powered tutoring systems, for instance. The potential to provide personalized, 24/7 learning support is undeniably exciting. But we must also consider the very human elements of education that cannot be replicated by algorithms. Our challenge is to find the sweet spot where emerging technologies amplify rather than replace the irreplaceable human connections in learning.

5. How to interpret and use the scenarios presented

Adopting the Future

The successful integration of technology in education extends beyond the traditional dichotomy of instructional approaches. Instead, effective technology integration hinges on systematic preparation and implementation strategies across all levels of educational leadership. This book explores how educational stakeholders, including classroom teachers, school administrators, district leaders, and policy makers, can best facilitate technology integration to develop effective digital learners prepared for a technology-rich workforce where digital competency is essential.

Theoretical Foundations

Three key theoretical frameworks inform our understanding of technology integration in education: technology integration theory, self-efficacy, and the diffusion of innovation model.

Technology Integration Framework

Don Ely's research on technological innovations provides a foundational understanding of implementation requirements. Successful implementation demands specialized knowledge for efficient and thorough execution. Ely's framework identifies eight critical variables for successful educational technology adoption:[24]

1. Dissatisfaction with current methods across all stakeholder levels

2. Existing knowledge and skills throughout the educational system

3. Resource availability and allocation

4. Time allocation for all educational professionals

5. Reward systems at institutional and policy levels

6. Participatory decision-making across hierarchies

7. Multi-level stakeholder commitment

8. Visible leadership from administration and policy makers

The *integration* of technology requires intentional leadership decisions addressing these variables. The *adoption* of technology however involves a more intimate understanding of the individual's choice to adopt.

Self-Efficacy in Educational Systems

Self-efficacy—one's perceived ability to succeed—plays a crucial role in technology adoption at all levels of educational leadership. According to Bandura, "Efficacy expectations determine how much effort people will expend and how long they will persist in the face of obstacles and aversive experiences."[25] This concept differs from actual achievement; it represents the belief in one's ability to execute necessary behaviors for success.

Research by Ghaith and Shaaban reveals that "Educational professionals who believe in their personal ability to provide effective implementation that would bring about student learning are less concerned about their self-survival and about the demands of the task than their less efficacious counterparts."[26] Their findings suggest that intervention programs focusing on increasing efficacy may be particularly beneficial for new educational professionals at all levels.

Innovation Diffusion in Education

Rogers' Diffusion of Innovation theory provides a framework for understanding technology adoption rates across educational systems.[27] Moore and Benbasat expanded this framework, adding three crucial attributes:

1. Voluntariness of adoption at various institutional levels

2. Impact on professional image across the educational hierarchy

3. Result demonstrability for all stakeholders

Their research offers us valuable insights for technology adoption across educational settings identifying the need to focus on individuals in the adoption process.[28]

Implementation Implications

Successful technology integration in K-12 education requires careful attention to stakeholder perception and support at all levels. Research indicates that prioritizing technology over curriculum creates only superficial connections between technological features and learning principles.[29] Instead, professional development and policy initiatives should focus on building institutional confidence and competence in integrating technology meaningfully into existing educational structures.

The success of educational technology adoption largely depends on each stakeholders' perceived ability to implement these tools effectively. Professional development programs and policy initiatives must address both technical competency and institutional self-efficacy to ensure successful adoption. This understanding suggests that future technology integration efforts should focus on building confidence alongside technical skills throughout the educational hierarchy, creating a more sustainable model for educational innovation.

Note: This approach acknowledges that while technology itself is important, the key to successful integration lies in comprehensive preparation and support across all educational levels, ensuring that technology serves learning objectives rather than driving them. Professional development programs and policy frameworks that recognize and address these factors will be better positioned to facilitate effective technology integration in K-12 education.

The Process versus The Product

There is an important new focus on the learning process. Many tech giants have come to understand that developing and releasing a technologically sound product with cutting edge innovation is simply not good enough. The customer needs, pain points, etc. should rule the product development process. With this in mind, we must design with the end-user, in our case the learner, in mind and constantly ask: What do they need to be successful in their future? This meaningful adoption narrative is historically grounded in efforts to integrate past technologies into education.

Adopting Innovation in Education: Lessons from Radio's Journey

The adoption of emergent technology in education has been a persistent challenge throughout history. By examining the journey of radio as an educational tool, we can gain valuable insights into the process of innovation adoption and apply these lessons to our current educational landscape.

The Fusion Model provides a framework for understanding the organizational adoption of innovation. As we explore radio's integration into education, we can identify the key stages of this model: Agenda Setting, Matching, Redefining/Restructuring, Clarifying, and Routinizing.

In the agenda setting stage, educational institutions in the 1920s recognized the potential of radio as a revolutionary tool for instruction. The vision articulated by Benjamin Darrow in 1932 exemplifies this stage: "The central and dominant aim of education by radio is to bring the world to the classroom, to make universally available the services of the finest teachers, the inspiration of the greatest leaders..."[30]

The matching stage saw schools attempting to fit radio technology into their existing structures. However, initial adoption was hindered by technological limitations such as poor battery life and reception, as well as uncertainty among educators about best practices.[31] These barriers mirror challenges we still face today with new technologies, highlighting the cyclical nature of innovation adoption.

The redefining/restructuring stage occurred as radio technology improved, particularly with the invention of the transistor in 1947. This advancement made radios more reliable, inexpensive, and portable, aligning with Moore's Law, which predicts the continuous improvement of technology over time.[32,33]

The clarifying stage was marked by educators grappling with how to effectively integrate radio into instruction. Initially, the one-way nature of radio communication posed challenges for interactive learning.[34] However, innovative educators developed strategies to overcome this limitation, such as creating supplementary materials and implementing interactive radio instruction (IRI) techniques.[35,36,37]

Finally, the routinizing stage was achieved as radio became a widely adopted educational tool. By 2001, radio had become the most important medium for development and social change worldwide, demonstrating its full integration into educational practices.[38]

This historical analysis reveals several key insights for adopting innovation in education:

1. Access is crucial but not sufficient: While technological improvements increased access to radio, professional development for educators was equally important in driving adoption.[39]

2. Adaptation is necessary: Educators had to adapt their teaching methods to leverage radio effectively, developing new instructional designs and supplementary materials.[40,41]

3. Cultural and organizational factors play a significant role: Pressure from parents and businesses influenced schools' decisions to adopt radio technology.[42]

4. Adoption takes time: It took approximately 25 years from radio's invention for its educational potential to be fully explored, a pattern that has repeated with other technologies like mobile devices.[43]

5. Innovation adoption is an active process: The shift from viewing radio as something to be integrated to something to be adopted represents a crucial change in mindset. Adoption implies empowerment and choice, while integration suggests a more passive acceptance.

As we face new technological innovations for the future, these lessons from radio's adoption can guide our approach. The Fusion Model reminds us that innovation adoption is a complex process involving multiple stakeholders and stages. By understanding this process, we can more effectively navigate the challenges and opportunities presented by emergent technologies in education.

Conclusion

To truly leverage innovation, we must focus on adoption rather than mere integration. This requires not only providing access to new technologies but also empowering educators to reimagine their instructional practices. As we continue to evolve our educational systems, we must remember that the most significant influence on innovation adoption is not always technological advancement, but rather the development of leaders, educators, and learners who can effectively harness these tools for improved learning outcomes.[44]

One hundred years or so ago, parents wanted to help schools ensure their children had access to the latest technology: the radio. Parents and businesses in the 1920s supplied schools across the United States with radio receivers in an effort to integrate trending technology into their children's educational experience.[45]

Yet simply providing the technology was not enough, educational content was lacking and took years to catch up.

Education is changing because the world is changing. Change is a bit scary due to unknown, unplanned consequences of our well-intentioned disruptions. The Fusion Model and our understanding on how people choose change can help guide us as we strive to provide a meaningful future in education that leverages the best tools and teaching practices to prepare our learners for what lies ahead.

Chapter 1: The Futurist's Toolkit - Conclusion

The Futurist's Toolkit

Frameworks and Models for Educational Innovation

♫ Fusion Model

Agenda Setting

Matching

Redefining / Restructuring

Clarifying

Routinizing

∿ Activity Theory

Subject (Individual/Group)

Tools & Instruments

Rules & Community

Division of Labor

☍ Innovation Champions

Innovators (2.5%)

Early Adopters (13.5%)

Early Majority (34%)

Late Majority (34%)

Laggards (16%)

⩘ Hype Cycle

Technology Trigger

Peak of Inflated Expectations

Trough of Disillusionment

Slope of Enlightenment

Plateau of Productivity

◉ Key Implementation Principles

Systematic Approach to Innovation

Understanding Adoption Patterns

Balancing Change with Stability

HORIZON SCAN

Understanding the future of education requires both methodological rigor and creative vision. Through the Fusion Model and careful analysis of innovation adoption patterns, we can better navigate the complex landscape of educational transformation. The tools and frameworks presented in this chapter—from scenario planning to the Hype Cycle—provide a foundation for thoughtful innovation adoption. As we move forward, these tools will help us balance technological advancement with human-centered learning, ensuring that our pursuit of innovation serves rather than overshadows our educational goals.

WAYPOINTS

Key Insight 1: Successful innovation adoption follows a predictable pattern that can be understood and managed through the Fusion Model's five stages: Agenda Setting, Matching, Redefining/Restructuring, Clarifying, and Routinizing

Key Insight 2: The power of an innovation lies not in naming it, but in its seamless integration into practice—achieving the "pencil moment" where technology becomes an unnamed part of our routine

Key Insight 3: Activity Theory provides a critical lens for understanding how different components of an educational system interact during innovation adoption

Key Insight 4: The Hype Cycle helps us anticipate and navigate the typical progression of emerging technologies, from initial excitement through disillusionment to productive implementation

NAVIGATION PROMPTS

⊙ Implementation

- How might you apply the Fusion Model to a current or planned innovation in your classroom or school?
- What strategies could help your school community persist when initial excitement about a new technology wanes?

💡 Innovation

- Which emerging educational technologies might be at different points on the Hype Cycle in your school or district?
- How might these tools transform teaching and learning in your educational setting?

🌏 Impact

- How might systematic innovation adoption affect different members of your school community (students, teachers, administrators, parents)?
- What role could Activity Theory play in understanding how new technologies influence classroom dynamics and learning outcomes?

FUTURE LOG

Observations:

Ideas to Explore:

Next Steps:

SCENARIOS
FOUR FUTURES FOR EDUCATION

In this chapter, we will unpack a future for education based on tech trends, the future of work, and societal demands. However, we must recognize this as a possible future, one that is subject to economic, societal, geographic, and political forces that will deter, and sometimes accelerate, its trajectory.

1. The Hyper-Connected Classroom (2029)

AI-driven personalized learning.

Global virtual classrooms and cultural exchange programs.

Challenges: digital divide, data privacy, and screen time concerns.

In this near-future scenario, we see the culmination of trends that are already emerging in 2024. The 2029 classroom is a blend of physical and virtual spaces, leveraging AI and global connectivity to create a rich, personalized learning environment.

AI in Education: a focus on pedagogy

AI in education is a rapidly growing field with the potential to revolutionize the way we teach and learn. AI will continue to be used to automate tasks, provide personalized feedback, and create immersive learning experiences. However, there are also a number of challenges that need to be addressed, such as data privacy and safety.

While AI is a valuable tool for teachers, helping them scale good practice and effectively reach more students, we need to proceed with care when deploying emergent technologies into schools, and keep data privacy as a fundamental concern that needs to be understood.

Assessment

AI tools will drive how we engage our students and provide opportunities to effectively and efficiently scale well-established instructional practice (pedagogy). Instructional assessment processes, that is, how we measure knowledge and skill acquisition, have always been a central component of education. Traditionally, educators have worked to balance summative assessment and formative assessment.

The 2029 classroom will challenge this tradition because the goal of summative assessment is to evaluate student learning at the end of instruction, often leveraging pre-existing benchmarks. This is an end-product, rather than process focused, area of assessment. What does this matter for AI? With a take-home essay, for example, can an educator truly trust who, or what, is doing the work? When responses were copy/pasted off of a website, educators learned how to flag the work but AI tools are increasingly providing unique responses that are challenging to identify.

The exciting contribution AI brings to the 2029 classroom is in the formative assessment realm. Here the goal is to monitor ongoing student learning and to provide meaningful feedback. It is difficult for an individual educator to give this type of ongoing feedback consistently for large groups of students. AI will be a powerful tool to personalize learning for students with rapid, in-the-moment feedback to nurture their comprehension during their learning process.

AI-driven, personalized learning becomes the norm. Each student has a digital learning profile that adapts in real-time based on their progress, preferences, and learning style. This builds on current adaptive learning platforms, but takes it several steps further. AI tutors provide instant feedback and adjust the curriculum on the fly, ensuring that each student is appropriately challenged and engaged.

Global virtual classrooms break down geographical barriers. Students regularly collaborate with peers from around the world on projects, fostering cultural understanding and global citizenship. This is an evolution of current learning management and video conferencing tools enhanced with seamless real-time translation and more immersive virtual reality experiences.

However, this hyper-connected world brings challenges. The digital divide becomes more pronounced, with students who lack access to high-speed internet or the latest devices falling behind. This void is increasingly filled by satellite connectivity.

Instruction

AI, like all technology, is an amplifier. In an educational setting, AI amplifies both good practice and strategies that need improvement. The 2029 classroom will rediscover the power of the apprenticeship model, arguably the most natural, effective form of teaching and learning, which has remained amongst modern humankind's greatest scalability challenges. We will disrupt our practice of filling auditoriums with eager learners, instead designing self-paced courses, and exploring countless feedback systems, all to replicate the painfully obvious: an individual, 1:1 *human touch* in learning is incredibly powerful but difficult to scale. In the 2029 classroom, educators will work individually with a learner to model, guide, provide feedback, challenge, and validate the learner's progress. Education will build off of the apprenticeship model having some or all of the following processes in place:[46,47,48]

- Modeling: The educator carries out the task, simultaneously thinking aloud about the process, while the learner observes and listens.
- Coaching: As the learner performs the task, the educator gives frequent suggestions, hints, and feedback.
- Scaffolding: The educator provides various forms of support for the learner, perhaps by simplifying the task, breaking it into smaller and more manageable components, or providing less complicated equipment.
- Articulation: The learner explains what they are doing and why, allowing the educator to examine the learner's knowledge, reasoning, and problem-solving strategies.
- Reflection: The educator asks the learner to discuss and compare their performance with that of experts, or perhaps with an ideal model of how the task should be done.
- Increasing complexity and diversity of tasks: As the learner gains greater proficiency, the educator presents more complex, challenging, and varied tasks to complete.
- Exploration: The educator encourages the learner to frame questions and problems on their own, and in doing so to expand and refine acquired skills.

While the above identifies a tried and true method of learning as an effective one on one scenario, historically proven a challenge when attempts are made to scale to a larger student audience. The 2029 classroom will make the apprenticeship model a highly effective pedagogical practice that leverages key components of social learning where profound meaning-making of individual learning is fully realized in a social, group setting. The new AI-driven apprenticeship model finds the educator modeling a task for their students, while explaining their thought process, then directing their students to a self-guided, AI-powered, series of tasks which are responsively-enabled to appropriately challenge the learner to achieve expertise (COACHING, SCAFFOLDING, ARTICULATION, REFLECTION, and INCREASING COMPLEXITY AND DIVERSITY OF TASKS).

Here the educator has been given the opportunity to work one on one with students requiring support beyond what the AI is offering. Finally, with developments in AI, students now have a new set of tools that they can use to produce AI generative products (ex. graphics and videos), via question prompts, that represent their content-area expertise (exploration). Throughout the 2026 classroom authentic discourse is intentionally planned as a learning imperative to humanize the process with educator-led check-ins and student-centric dialogue.

Further, we will hear of innovative educators modeling AI applications through transparent examples with their students, identifying the pros and cons of the technology as it exists in 2029. The modeling approach, where the educator speaks openly about how they use AI, and even explain their reservations, is a life skill lesson about having grit, or agency, to persevere and overcome a new challenge. By demonstrating a level of stick-to-it-tiveness that we all need in life, educators will discuss their use of AI as an example that provides opportunities for their students to have clear access to the skill of *agency* in action.

Safety and Security

The learning experience must always be humanized, AI is not a replacement for teachers. Teachers are essential for providing students with the guidance and support that they need to succeed. AI can be a valuable tool for teachers, helping them scale good practice and effectively reach more students. But we need to proceed with care when deploying emergent technologies into schools. We need to be appropriately hyper-sensitive on how these technologies will impact our students. Regarding AI, the pedagogy will get worked out, teachers are masters of change, that is not to undermine the significance of the instructional shift, but it can be done. The 2029 challenge we need to resolve is centered on safety and security. Data privacy is a fundamental concern that has to be resolved and better understood. Learners need to ask questions, work through problems, and create less than perfect solutions, all of which are not meant for the public stage, their privacy matters and must be protected.

In 2029 data privacy concerns will have reached a fever pitch as parents and educators grapple with the implications of AI systems having access to detailed learning and behavioral data on students. Further there are growing concerns about screen time and its impact on physical and mental health. Wearable technology in multiple forms, including smart watches and AR Glasses will exacerbate this concern with an increasingly always-online, connected lifestyle.

Change is uncomfortable and sometimes scary but change,

whether forced (ex. COVID teaching), or by choice, has the potential to guide us to make the world a better place for our children.

2. The Bio-Integrated Learner (2039)

Neural implants for knowledge acquisition.

Ethical debates on cognitive enhancement.

Implications for assessment and credentialing.

By 2039, AI will no longer be thought of as a tool independent of learning. We will not say we are "learning/working with AI" rather, full adoption will have occurred. The integration of technology and biology will have progressed to a point where neural implants for knowledge acquisition become a reality. This represents a significant leap from the non-invasive interfaces and wearables we see in development today. Yet, like most innovation the old exists while the new is being adopted. Non–invasive interfaces and wearables will becoming increasingly responsive to brain activity even leveraging AI to complete an action from an inference.

More invasive implants will download information directly to a device that is accessible through a person's brain, dramatically accelerating the learning process for certain types of knowledge. Language acquisition, historical facts, and scientific principles can be assimilated in a fraction of the time it takes through traditional learning methods.

This technology sparks intense ethical debates that transcend screen time and authenticity of authorship, and also amplify mental health concerns. Questions of fairness arise as those with access to these implants gain significant advantages. A new narrative about equitable access will occur and rapidly make our smartphones-in-the-classroom dilemma seem like ancient news. There are concerns about the long-term effects on brain plasticity and the potential loss of critical thinking skills if information is simply "downloaded" rather than learned through experience and reflection.

The implications for assessment and credentialing are profound. Traditional exams become obsolete for measuring factual knowledge. Instead, education shifts towards evaluating how students apply knowledge, their creativity, and their ability to synthesize information from various sources.

3. The Community Learning Hub (2049)

Decentralized, project-based learning centers.

Integration of education with local industry and community needs.

Shift from age-based to competency-based progression.

By mid-century (2049), we see a radical restructuring of the educational system. The traditional school building evolves into a community learning hub, reflecting a shift towards more decentralized, project-based learning.

These hubs are deeply integrated with local industry and community needs. Students of various ages work together on real-world projects that benefit their communities, guided by educators who act more as facilitators than traditional teachers. This approach builds on the current trend towards project-based learning and soft skills development, but takes it to a new level of community integration.

The concept of age-based grades disappears, replaced by a

competency-based progression system. Students move through learning modules at their own pace, demonstrating mastery through practical applications rather than standardized tests. Preparing for either a gig-based role in the globalized economy or a hyper-local contributor role—or perhaps both.

This model addresses many of the criticisms of the traditional education system, such as its disconnect from real-world skills and its one-size-fits-all approach. However, it also presents challenges in terms of standardization and ensuring equitable access to learning opportunities across different communities.

4. The Post-Scarcity Scholar (2059)

Universal access to all human knowledge.

AI tutors and holographic expert projections.

Redefinition of the purpose of education in a highly automated world.

By 2059, we enter a world where access to information is no longer a limiting factor in education. Universal access to all human knowledge is a reality, building on trends we see today with open educational resources and massive online repositories of information.

AI tutors become indistinguishable from human teachers in many respects, fueling debates over what "humanizing" learning really means. AI can take on any form—from a text interface to a fully realized holographic projection—and can access the sum total of human knowledge instantaneously. Expert knowledge on any subject is available on-demand through holographic projections, allowing students to interact with simulations of the greatest minds in history.

In this world of abundance, the purpose of education is radically redefined. With AI handling most routine cognitive tasks, education focuses on uniquely human skills: discernment, creativity, emotional intelligence, ethical reasoning, and the ability to ask good questions. The role of human educators shifts towards mentorship, helping students navigate the vast sea of available knowledge and experiences.

This scenario presents its own challenges. With information so readily available, the value of traditional credentials is questioned. The education system must grapple with how to prepare students for a world where many traditional careers have been automated, and where the pace of change is faster than ever before.

Conclusion

These scenarios paint a picture of an education system that evolves dramatically over the next 50 years. While technology plays a crucial role in each scenario, the fundamental human aspects of learning— curiosity, social interaction, and the joy of discovery—remain central to the educational experience. As we move forward, the challenge will be to harness these technological advancements in ways that enhance rather than replace the human elements of education. As we will see, many tenets of good teaching must be retained. The future will need us to continue to humanize the learning experience.

Chapter 2: Scenarios - Four Futures for Education - Conclusion

Four Futures for Education

Scenarios and Transformations Towards 2059

 **Hyper-Connected Classroom** (2029)

AI-Driven Personalized Learning

Global Virtual Classrooms

Cultural Exchange Programs

Digital Privacy Considerations

Bio-Integrated Learner (2039)

Neural Implants for Learning

Cognitive Enhancement Ethics

New Assessment Methods

Equity Considerations

Community Learning Hub (2049)

Decentralized Learning Centers

Industry-Education Integration

Competency-Based Progress

Local Community Focus

Post-Scarcity Scholar (2059)

Universal Knowledge Access

AI Tutors & Holographic Experts

Redefined Purpose of Education

Post-Automation Learning

→ **Key Transformations**

| Technology Integration | Human Enhancement | Community Focus | Knowledge Democratization |

HORIZON SCAN

Our exploration of possible educational futures reveals both exciting opportunities and significant challenges. From the hyper-connected classroom of 2029 to the post-scarcity scholar of 2059, each scenario presents a unique vision of how education might evolve. These futures are not mutually exclusive—elements of each may emerge as we progress through the coming decades. By examining these potential futures now, we can better prepare for the challenges and opportunities they present, ensuring that our educational systems remain both innovative and human-centered.

WAYPOINTS

Key Insight 1: The progression from hyper-connected classrooms to bio-integrated learning represents a fundamental shift in how we think about knowledge acquisition

Key Insight 2: Community learning hubs and decentralized education models may become increasingly important as traditional institutional boundaries blur

Key Insight 3: The evolution of AI tutors and holographic experts will require careful consideration of the balance between technological efficiency and human connection

Key Insight 4: Universal access to knowledge will require us to rethink the fundamental purpose and methods of education

NAVIGATION PROMPTS

🎯 Implementation

- How might your institution begin preparing for the transition to more personalized, AI-driven learning?
- What steps could you take to strengthen community connections in anticipation of more decentralized learning models?

💡 Innovation

- Which elements of these future scenarios could be beneficial to implement in your current context?
- How might you balance the promise of new technologies with the need for human connection?

🌍 Impact

- How might different stakeholders in your community respond to these potential futures?
- What steps could ensure that technological advancement promotes rather than hinders educational equity?

FUTURE LOG

Observations:

Ideas to Explore:

Next Steps:

PEDAGOGY AND THE FUTURE

As we seek to understand the future of education we must reflect on the methods we employ on a daily basis to learn something new or reinforce previous learnings to be successful in our tasks. Many of these methods are tried and true with a solid foothold in our future.

Think about the process of learning to ride a bicycle. We might be highly motivated to ride it, read and research about how to do it, and ask for expert advice, but until we actually get on the bicycle and try, we do not have the skills needed to ride it successfully. However, our motivation and our depth of knowledge will help us to safely take advantage of all that bicycle riding has to offer us for the rest of our lives.

The power of storytelling, the coach or guide who led you to success, the compassionate individual who humanized your learning and sometimes life circumstances, these are the examples of good teaching practice. As we unpack the future of education it is critical that we acknowledge many aspects of effective pedagogy that, at their core, emphasize what it is to be human.

Pedagogy can be considered the backwards design of strategies that are meant to address how we learn best. What follows is not an exhaustive review of education pedagogy, rather a selection of pedagogy that I believe will heavily inform the future of education. Further research on the nuances of educational strategies and theories will benefit readers seeking to understand more about the "why" of these selections.

1. How We Learn

Human Performance Technology

Human performance technology is a systematic approach to improving the productivity and competence of people in organizations. Rooted in behavioral psychology, this field of practice helps us understand four attributes that, if identified and fostered, will lead to success: knowledge, skills, environment, and motivation.[49] Knowledge is what we know through traditional learning settings that establish the contextual "why" of concepts. Skills are gained through applied training and practicing the "what" of knowledge. Environment describes conditions in which we learn, work, and live. Motivation is both our intrinsic and extrinsic foci that drive us.

If we do not take care to nurture these attributes, we will have an experience rife with anxiety, frustration, limitation, and/or apathy.

Within the context of this book we understand the attributes as fundamental players in productivity and competency in educational institutions.

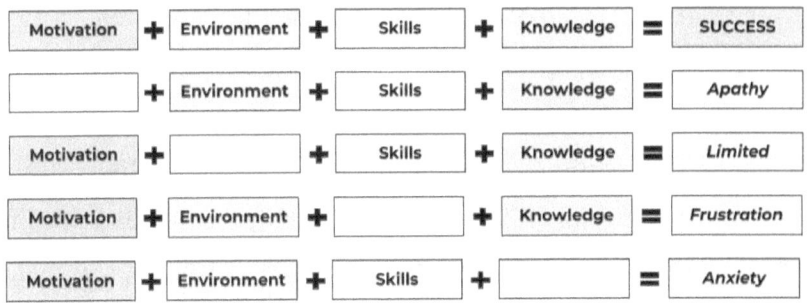

From: WanderlustEDU *Chapter 2: The Journey to Transform School Culture Starts with Teachers.*

Motivation

Motivation can be categorized two ways: extrinsic and intrinsic. Extrinsic motivation comes from outside of each of us and includes: salary, tenure, retirement, and healthcare. Intrinsic motivation comes from inside of us, it keeps us in difficult times and drives us to excel at what we do. Intrinsic value of the work we do each day is inherently human which sustains us through difficulties and provides a true sense of accomplishment.

On the other hand, intrinsic motivation arises from within an individual and is fueled by personal interests, values, and a sense of purpose. It is the inherent satisfaction and enjoyment derived from the work itself, independent of any external rewards. Intrinsic motivation is deeply connected to an individual's sense of autonomy, mastery, and purpose. When individuals feel that they have control over their work, are able to develop their skills and expertise, and find meaning in what they do, they are more likely to be intrinsically motivated. This type of motivation is often associated with greater creativity, innovation, and long-term engagement.

The intrinsic value of work is closely tied to our human nature and our desire to contribute to something meaningful. It is this sense of purpose that sustains us through challenges and setbacks, and provides a true sense of accomplishment. When we are intrinsically motivated, we are more likely to persevere in the face of difficulties, to seek out new challenges, and to continuously strive for improvement. Intrinsic motivation is not only crucial for individual well-being and fulfillment but also essential for organizational success.

While extrinsic motivation can be effective in certain situations, it is ultimately intrinsic motivation that drives individuals to excel and reach their full potential. Recognizing and fostering intrinsic motivation in the workplace is essential for creating a sustainable and thriving work environment where individuals are not only productive but also engaged, satisfied, and committed to their work.

Environment

The impact the environment has on the learning process is undeniable.

Research in this area points major trends, top among them are:[50]

1. Space is only one factor.

2. Perception is reality.

3. Involve users in design.

4. Teacher professional development is critical.

1. Space is only one factor.

Most research is on relations between design, learning activities and learning results. Space cannot be isolated as a single cause to positive learning outcomes, but people, space, interaction and learning are intertwined.[51]

2. Perception is reality.

Closely connected is the theme on how space is perceived by teachers and students. Perception of space is emotional but also intertwined with the pedagogy used in the space.[52]

3. Involve users in design.

Only when a learning environment is seen to support learning and create a positive experience can we say it is designed successfully.[53] While researchers explore different aspects of the design process, design principles and participatory design projects, they largely agree on one key point: the need to involve educators early in the design process.[54, 55]

4. Teacher professional development is critical.

There are very few articles on how teachers are supported in their use of space. None of the articles look for evidence on the best way to support teachers, but highlight that support is needed.[56]

Further, the assessment of how meaningful the impact the environment specifically has on learning is challenging. The literature on methods to evaluate the complex relationship between learning spaces and student learning is both scarce and fragmented.[57] Most articles have no explicit theoretical perspective, but in the few that have, there is a movement towards an understanding of space as being produced in an entanglement of people, social and material resources and that many different theoretical perspectives are used to frame this understanding.[58] Sorensen[59] and Mulcahy[60] have both found a socio-material approach to be meaningful in understanding how people, space, and practice are interconnected. Their research reveals how physical environments actively shape learning processes rather than serving as passive backdrops. Sorensen's work particularly demonstrates how technological tools and spatial arrangements mediate knowledge construction, while Mulcahy's analysis of museums and schools highlights how learning spaces emerge through the dynamic interplay of human actors, material objects, and educational practices. Together, these studies underscore that effective learning environment design must consider not just the physical layout, but the complex web of relationships between learners, educators, materials, and spatial configurations that collectively produce meaningful educational experiences.

Skills

Skill acquisition is an iterative process, experiential application of skills will lead to expertise. Skills are directly related to the emergence of new technology in classrooms. Think about the fact that the most advanced piece of technology you use each day is the lowest piece of technology our students will use in the future. Technology regularly causes disruption in education; it forces us to exist in a state of constant change, a state of innovativeness. It can be intimidating to try to apply every new skill we learn.

I think of the earliest explorers of the earth, who were told the earth was round. Scientists and cartographers explained that the earth was round and demonstrated it with stunning visual aids. These great thinkers gave direction and contributed to the planning of brave explorers, but not many of them went along to test their theory, to prove their point. So is the educator's journey, we are told new innovations or best practices will work, but it's up to us to bravely go forth and apply what we learn.

Gaming provides us with an excellent example of gaining and applying new skills. This form of growth is highly disruptive to our modus operandi, we must stretch our comfort zone to increase our skills.

Games (and simulations) are an important way to create meaningful, experiential knowledge for students. Games allow us to fail, they also allow us to see the benefit in learning new skills through experience. A fail-forward mentality is a skill set necessary for our students. For our learners to master grit, a sense of stick-to-itiveness in adversity, they must observe it modeled by adults who attempt new things with a certain degree of failure.

Games can capture the attention of our students, can be relevant to their world, appeal as an approachable challenge, and offer frequent feedback, leading to student high level of engagement satisfaction. When employing gaming in the classroom, it is the work of the educator to relate and connect the activities to the required content and at times to critically teach learners to examine themes and representations of people that are unfair and/or unjust. Admittedly, there have been times in my own classroom where I got caught up in the student excitement over gameplay and let the content-learning take care of itself, which it seldom does.

The positive student reaction to gaming experiences in the classroom is both powerful and transformative. If you are thinking of taking on a little more advanced gameplay like the example discussed here, I highly suggest user testing. Start a club or find some kids in a study hall and have them work out the kinks for you, they will love it, and you will have meaningful data to inform a full class rollout.

It is so important for educators to embrace an agency mindset and innovate to support students. But to innovate means to accept the possibility of failure, believe it or not, it is a good thing for students to see adults fail and to problem solve. Telling students to be resilient, to persevere, to struggle through difficulties is good, demonstrating it is better.

Knowledge

Educators develop knowledge retention strategies, like skill acquisition, through iteration and strategic planning. Traditionally, educators studied their specific content areas with the intent of being the content-expert, the sage-on-the-stage, the one who knew all the answers. A paradigm shift to best serve learner needs as a guide-on-the-side has positioned educators as 1:1 coaches whenever possible. Due to the power of both storytelling and 1:1 coaching, the greatest practice is a mix of both, much like the classical apprenticeship model. The most powerful form of knowledge acquisition is through experience—this knowledge can, at times, appear to contradict our training, that is, allowing failure to occur in a safe space.

Knowledge retention is naturally achieved through experience and with that understanding I would ask that you consider the agency mindset, a gamer mentality, where we believe we have the capacity to take on a challenge. We may not all be gamers or enjoy video games, but we all can learn so much from going and doing, from failing, and from success. And at each failure we should change our method, doing the same thing over and over again, even if well intentioned, is just plain insanity. The difference between an expert and beginner is that the expert has failed more times than a beginner has even tried.

2. How we choose to adopt innovation

Each time we learn something new we think, "Oh! It's just like when I do this other thing." Learning to play chess you might think about how the board looks *like* a checkerboard (if you have played checkers). You then understand how the play pieces are meant to operate in a like manner, that is, inside the boxes. That "like" relationship is called our schema.

Schemas and past experiences play crucial roles in decision-making by providing mental frameworks that help individuals organize and interpret information. Schemas are cognitive structures that allow people to simplify complex information and make quick decisions with minimal cognitive effort.[61, 62] They are built from past experiences and knowledge, which help individuals predict outcomes and make informed choices.[63, 64, 65]

When faced with a decision, people often rely on their schemas to assess potential outcomes and risks.[66] This leads us to efficient decision-making since we have a basis for interpreting new information and integrating it with existing knowledge.[67, 68] However, schemas can also limit decision-making by causing individuals to ignore or reject information that does not fit their existing frameworks, leading to confirmation bias.[69, 70, 71] Basically if we can relate to something new we are more likely to choose to adopt it.

Microsoft Zune sold portable MP3 players boasting speed and storage yet Apple sold 10,000 songs in your pocket with the iPod. People related more to records than MP3s, Apple understood the power of schema and relatability.[72]

Past experiences contribute to the formation and evolution of schemas, influencing how new situations are perceived and interpreted.[73] These experiences can either reinforce existing schemas or lead to their modification when new information challenges them.[74] Overall, schemas and past experiences are integral to decision-making, affecting both the efficiency and accuracy of the process.

As we seek to understand the future it is critical that we acknowledge our past experiences as shaping our preferences and biases about what might be.

3. Pedagogy

In a rapidly changing world literacy is a baseline for ensuring future generations are critical thinkers and active participants in creating a better future. We must emphasize critical review of content as a key component of literacy, literacy skills are the currency of the information age.[75]

A traditional model of teaching discussed previously, assumes all learners will progress at nearly the same rate in acquiring knowledge and skills yet most of us understand this as an imperfect assumption for educational systems. The upskilling/reskilling efforts in the workforce promote a different model, a competency-based model. This is important to understand since education is designed to prepare students for their future. If the future of work involves a need to regularly learn new skills for work, education should teach the skill of self-learning.

Competency-based learning is an educational approach that focuses on gaining expertise in specific skills or competencies rather than simply spending a set amount of time studying a subject. Here are the key aspects:

1. Skill focus: The curriculum is organized around specific competencies or skills students need to demonstrate.

2. Self-paced: Students progress at their own pace, moving on only when they've mastered a competency.

3. Personalized: Learning is tailored to individual student needs and abilities.

4. Outcome-oriented: The emphasis is on demonstrating mastery rather than completing a certain number of hours or courses.

5. Flexible assessment: Students are assessed when they feel ready, not on a fixed schedule.

6. Real-world relevance: Competencies are often tied to skills needed in professional settings.

7. Clear expectations: Students know exactly what they need to learn and demonstrate to progress.

This approach contrasts with traditional time-based models ubiquitously found throughout the education space where all students move through material at the same pace regardless of individual mastery. But competency-based learning provides only one side of the learning journey, social-emotional wellness and development are critical.

Socio-emotional learning and whole-child approaches

Research on mental health states Gen Z and young millennial employees are missing the equivalent of one day's work every week due to mental health concerns.[76] In an always-online, albeit hyper connected world we must prepare our students to balance in-person and digital connections in order to manage one area of mental health and better prepare them for their future.

Social Learning Theory

Social learning theory demonstrates a dynamic evolution across different life stages, with distinct characteristics and impacts at each developmental phase. In childhood, individuals exhibit a strong propensity for imitation and modeling behaviors, laying the foundational groundwork for skill development and attitude formation through observational learning.[77, 78, 79, 80, 81, 82] As individuals transition into adolescence, the focus shifts dramatically toward peer influence,[83,84] where social learning becomes instrumental in shaping identity, self-concept, and understanding of social norms, with teenagers becoming increasingly selective in their imitation based on their desired self-image.[85, 86, 87] Adulthood marks another significant shift, characterized by more discriminating social learning patterns influenced by accumulated experiences and values,[88, 89] particularly in professional contexts,[90, 91] where individuals typically demonstrate a more critical approach and reduced susceptibility to peer influence.[92, 93]

Throughout these stages, the effectiveness of social learning remains intrinsically linked to both cognitive abilities and social development, with each factor playing a crucial role in determining how successfully individuals can observe, process, and implement learned behaviors and knowledge.[94, 95, 96, 97]

This age-dependent progression in social learning highlights its fundamental role in human development, from the simple imitation of childhood to the complex, selective learning processes of adulthood.

Social Learning and Play

The interplay between social learning[98] and play[99,100] in educational development reveals both striking parallels and notable distinctions. While both processes emphasize the fundamental importance of observation, imitation, and social interaction as learning mechanisms that evolve with developmental stages, they diverge significantly in their underlying frameworks and approaches.

Social learning theory centers on the structured acquisition of behaviors, attitudes, and values through observational learning and modeling, grounded in principles of reinforcement and cognitive processing. In contrast, play-based research, drawing from developmental theories like those of Piaget[101] and Vygotsky[102], explores how various forms of play contribute to growth through active engagement and spontaneous creativity. This theoretical distinction manifests practically in how each process unfolds—social learning typically follows more structured patterns of observation and modeling, while play embraces spontaneity, imagination, and hands-on exploration.

Understanding these similarities and differences helps illuminate how both processes work together in supporting comprehensive learning and development, suggesting that effective educational approaches might benefit from incorporating elements of both structured social learning and free-form play.

Conclusion

As we conclude our exploration of pedagogy and its future implications, several enduring truths emerge about the nature of learning and teaching. While technology and educational methods continue to evolve, the fundamental human elements of learning remain constant. The four pillars of human performance technology—knowledge, skills, environment, and motivation—provide a framework not just for understanding how we learn, but for anticipating how education must adapt to meet future challenges.

Our examination reveals that successful learning environments of the future will need to balance traditional human elements with technological innovation. The power of storytelling, personalized coaching, and compassionate guidance cannot be replaced by technology alone, but rather must be enhanced by it. This is particularly evident in how motivation shapes learning outcomes, with the interplay between intrinsic and extrinsic factors becoming increasingly important in an age of rapid technological change.

The shift from traditional "sage on stage" approaches to more collaborative "guide on side" methodologies signals a broader transformation in education. This evolution reflects our growing understanding of how skills are acquired through iteration and experience, how knowledge is built through active engagement rather than passive reception, and how environmental factors shape learning in complex and interconnected ways.

Perhaps most significantly, our exploration of schemas and their role in innovation adoption points to the critical importance of understanding how both educators and learners approach change. As we move forward to examine the factors that drive educational transformation, these foundational elements of pedagogy will serve as crucial reference points for understanding both the opportunities and challenges that lie ahead.

The future of education will not be shaped by choosing between human elements and technological advancement, but by thoughtfully integrating both to create learning experiences that are more engaging, effective, and equitable than ever before. This understanding will be essential as we delve deeper into the factors that catalyze and constrain educational change in our rapidly evolving world.

Chapter 3: Pedagogy and The Future - Conclusion

Pedagogy and The Future

Key Components of Modern Learning

Human Performance Technology

Knowledge Development

Skills Enhancement

Environmental Factors

Motivation Elements

Social Learning

Peer Interaction

Collaborative Learning

Cultural Context

Group Dynamics

Learning Environment

Physical Space Design

Digital Integration

Resource Accessibility

Safety & Comfort

Future Focus

Adaptive Learning

Technology Integration

Personalized Pathways

Innovation Adoption

Key Takeaways

Integration of Technology and Human Elements

Balanced Approach to Innovation

Focus on Continuous Adaptation

HORIZON SCAN

As education evolves, pedagogical foundations remain crucial even as delivery methods transform. The integration of technology must enhance rather than replace effective teaching practices, with particular attention to human performance technology, social-emotional learning, and the development of agency in both educators and students. Success in future education requires balancing innovation with proven pedagogical principles while maintaining focus on the human elements that make learning meaningful.

WAYPOINTS

Key Insight 1: Effective pedagogy in the future will blend traditional teaching wisdom with new technological capabilities

Key Insight 2: Human Performance Technology provides a framework for understanding and optimizing learning environments

Key Insight 3: Social-emotional learning and whole-child approaches become increasingly important as technology advances

Key Insight 4: The development of agency in both educators and students is crucial for future success

NAVIGATION PROMPTS

🎯 Implementation

- How might schools balance traditional pedagogical approaches with emerging technologies?
- What aspects of Human Performance Technology could strengthen teaching practices in your school community?

💡 Innovation

- How could new technologies enhance rather than replace the human elements of education?
- What opportunities exist for implementing competency-based learning across different grade levels?

🌐 Impact

- How might shifts in pedagogical approaches affect various stakeholders in your educational community?
- What role could social-emotional learning play in preparing school communities for the future?

FUTURE LOG

Observations:

Ideas to Explore:

Next Steps:

THE DRIVERS OF CHANGE

FORCES SHAPING THE FUTURE OF EDUCATION

1. Technology Revolutions

Fourth Industrial Revolution

Beyond the 4C's: Framing our Understanding of Future Skills in the era of the Fourth Industrial Revolution

In 2002, the National Education Association (NEA) formed the Partnership for 21st Century Skills (P21). P21 developed the "Framework for 21st Century Learning," which highlighted 18 essential skills for learning in the 21st century. However, it became evident that 18 skills were too broad, so in 2004, the "Four Cs" were introduced: Critical Thinking, Communication, Collaboration, and Creativity and Innovation.[103,104] For a frame of reference, 2004 was when Facebook launched. Needless to say our society has changed dramatically since then, now 20 years later we must revisit our understanding of what our students need for the next 20 years and beyond.

One way of framing our reflection is through the context of Industrial Revolution eras. These eras strive to organize the historical processes of change from an agrarian and handicraft economy to one dominated by industry and machine manufacturing and beyond.[105] The changes innovations have brought to us have introduced novel ways of working and living and fundamentally transformed our society. We an loosely define these eras as:[106]

Era	Century	Realization
First	18th Century	Mechanization: steam, water, mechanical production equipment
Second	19th century	Mass Production: division of labor, electricity, mass production, assembly line
Third	20th Century	Automation: electronics, computers, automated production
Fourth	21st Century	IOT and AI: cyber-physical systems

We now find ourselves well into the Fourth Industrial Revolution (4IR) characterized by the convergence of digital, biological, and physical technologies, which is creating new opportunities and challenges for businesses and society. One of the most significant impacts of 4IR is the rise of artificial intelligence (AI). AI is already being used in a variety of ways, from automating tasks to powering self-driving cars. As AI continues to develop, it will have an even greater impact on our lives and work. How then do we begin to reframe our thinking about education?

While our teaching and learning environments are changing, our motivation to prepare students for a successful future remains steadfast. Our perspective on the knowledge and skills required for success will become more clear as we struggle through necessary iterations of our thinking. Knowledge for example, as framed by curriculum, must increasingly be more globalized in its objectives and less hyper-focused on regional needs.

The very nature of an educational curriculum is to serve as a dynamic, iterative cultural artifact, that is, a reflection of what our society deems important. Historically, educational curricular shifts have gone from Socratic dialogue, to expert-apprentice programs, to humanities enrichment, to STEM focus, each with varying degrees of equitable access. Gaps in equity have led to economic-driven curricular designs (ex. STEM/STEAM). With ever increasing access to learning we must refocus our curricular priorities to address the needs of our global society from our positions as educators, administrators, policymakers, and parents. With this in mind we need a globalized curriculum that is culturally responsive and iterative in nature seeking the good of humanity over governmental initiatives.[107]

What are the skills of the future workplace? How are we to prepare our students now in 4IR? As discussed, we often talk about future-ready skills as related to the 4C's

1. Communication

2. Collaboration

3. Critical Thinking

4. Creativity

4IR skill demand requires an unpacking and reframing of these skills to appear more focused on growing learners in upskilling, intuitive decision making, rational decision making, collaborative innovation, and societal systems—all of which we will discuss later on in this chapter.[108]

The conversation about knowledge and skill development is never settled, nor should it be. We must continually ask questions about the status quo and challenge ourselves to prepare our learners for a successful future.

Artificial Intelligence and Machine Learning

AI's impact on education is still being defined but the change coming for educators and students is inevitable. An AI-driven, job market shift will have a profound impact on our educational systems as we prepare students for a changing world. Forty percent of all US work activity can be augmented, automated, or reinvented right now with AI.[109] These changes will impact many different sectors but perhaps one area to focus on is low wage jobs which includes manufacturing, food, hospitality and agriculture.[110] Many of these type jobs will no longer exist, leaving an on-the-job learning gap that educational systems will need to fill. Further, many of these jobs can be seen as temporary for younger generations seeking to enter the workforce. The sense of responsibility, customer service, and problem solving learned in these roles will now fall on schools to teach as part of a stronger, human-centric curriculum focus that skill development centering on working with AI.

The US Department of Education's Educational Technology Department stated: "We envision a technology-enhanced future more like an electric bike and less like robot vacuums. On an electric bike, the human is fully aware and fully in control, but their burden is less, and their effort is multiplied by a complementary technological enhancement."[111] Here our students must learn to co-create and collaborate with AI technology.

When we take a step back from all the hype, we see that AI is everywhere and that it has been around for some time. Dating back to 2017, Samsung's digital assistant, Bixby, lets you get things done using your voice. Bixby can send texts, check the weather, call your friends and family, launch apps, and control music playback like other voice assistants. Digital assistants are now leveraging AI to do more than simply recognize our voice commands, they now can suggest and further empower our workflows.

This will lead to formerly unheard of jobs like being a prompt engineer. This is a technique used in AI to optimize and fine-tune language models for particular tasks and desired outputs. Also known as prompt design, it refers to the process of carefully constructing prompts or inputs for AI models to enhance their performance on specific tasks. Our students' future will be full of thoughtfully crafted prompts to solve global issues. What a thought!

Yet, 15 years ago not many of us were thinking about a social media influencer as a job! But we understood the importance of Web 2.0 and our need to provide opportunities for our students to grow in this area. We learned to transition our online experiences from consumer of content to creator, a shift from static to dynamic that reflected in our classrooms. We did it then, we can do it now.

Forming succinct curriculum and designing instructional interventions centering on AI is a daunting challenge. Thinking about past innovations such as radio, computers, YouTube, and interactive displays, we find ourselves struggling to find the best way forward.

History has taught us that our process to adopt innovation is predictable. When we can relate to past experience we can find a positive way forward. In education, we experienced this at Samsung, we know adoption of innovation is nothing new and we have leveraged our experience in the space to design products like interactive displays to meet the current needs of educators, while remaining true to their experience. We designed our tools to offer simplicity and relatability for educators. While chalkboards, whiteboards, overhead projectors, and LCD projectors provided educators with a means to share examples and visuals for students. It was not until interactive displays were made available that now we could not only display content but illustrate and highlight visuals in a more meaningful way.

Every disruptive educational innovation from chalk, to the pencil, to 1:1 computing devices, prompted the same educational practitioner to ask the same questions over and over again:

- What will I do with it?
- What will my students do with it?
- How will I monitor their use?
- How will I maintain it?

In our previous exploration of School 1234 we discussed the organization, that is school wide, adoption of innovation. In reality, however, an "organization" might be a school district, a school building, or even a classroom—the adoption of innovation principles apply at every level. Here with each new innovation, we can test, recheck, tryout tools and methods to see what works. These are integration-focused efforts in the initial phases of adoption (Agenda-Setting and Matching). In these phases we seek to understand how tools, like AI, might be useful and meet our needs. Here, most of the experiments are within the bounds of the school's policies but even if you are locked down, be openly reflective of how it impacts everyday life. Talk to your students about it! Understand their workflows and yours. Here we can continue to give our students access to our adoption process, including the struggles. We need to give them access to agency.

Agency is about having a desire, making plans, and carrying out actions. The sense of agency plays a pivotal role in cognitive development, and includes the ability to recognize oneself as the agent of a behavior and as an empowered independent person. Agency is defined as the capacity of individuals to act independently and to make their own free choices. Agency is one's independent capability or ability to act on one's will. Each of us exercises the power of agency when we look up how to do things. We use YouTube videos to figure out how to repair a refrigerator or dishwasher. This is us embracing our own sense of agency.

Agency is the attitude that you can figure things out. Psychologist's work in positive perceived self-efficacy sheds light on the concept of agency by illustrating that we can *learn* to not be concerned with the number of skills that we have, but with what we believe we can do with what we have, under a variety of circumstances. Teaching with emergent technology is *not* about teaching with emergent technology. We need to guide our students to a sense of agency by providing them with access to our own agency in action.

As we move to the Implementation phases (Redefining/Restructuring, Clarifying, and Routinizing) our focus is on adoption of innovation (versus simple integration). We will see how to meaningfully shift our practice to better meet our learners' needs and prepare them for their future.

AI and the rise of Editors: Refocusing Educational Assessment in preparing our learners for their future

Assessing knowledge and skill acquisition has been a top priority in every form of education from early philosophers questioning their audience, to master craftsman coaching their apprentices, to today's formalized education systems. As we know, the Industrial Revolution had a profound impact on the later drive for summative, or end-product, deliverables to demonstrate mastery. This focus treats the learners as a product themselves, that is, a worker crafted to deliver predefined end results. This was industry's attempt at creating a viable workforce, to satisfy societal demand, during what we might call the first three Industrial Revolution eras, yet now we are in a fourth:[112,113]

In our Fourth Industrial Revolution era, our workforce needs have dramatically shifted, which should inform our educational assessment practices. Here we find our workforce co-collaborating with technology as end results are produced. The new skills for this workforce are significantly more process- than product-oriented, demanding a formative assessment focus on the learning process to help foster critical thinking and a growth mindset.[114] The end result will prepare our learners to be editors of products rather than widget/cog focused assembly line workers thinking solely about the product in front of them. The Fourth Industrial Revolution will lead to the rise of editors, are we ready?

Dropping AI: Seeking our Pencil Moment for AI in education

aiEducation?

aiLearning?

aiLesson?

aiTool?

aiSchool?

What will it take for AI to become so fully adopted that it is an unnamed part of our routine? Will AI have its pencil moment?

At some point over 100 years ago teachers were saying, "Today we are going to learn math with our pencils." Math *with* our pencils? Would we have it any other way today?

It is fascinating to think of a classical school tool, the pencil, as an innovation and spoken about as independent of the normal learning process: learning with a pencil. But at some point we dropped the "with a pencil" part of the statement and now we simply say, "Today we are going to learn math." We do not need to include naming the pencil, the innovation, which became part of our routine in learning. Each innovation, if meaningfully adopted, is destined to have a pencil moment.

Think about it: The real power in an innovation is not in naming it, rather, it is in not having to name it at all! Or that it becomes the Kleenex-brand, a brand originating product name for facial tissues. How can this be the case? A pencil moment occurs when an innovation not only solves a problem (the *what*) but it does so effectively because it is powered by its relation to the *why* (a question of practice).

The post-pandemic education prefix of choice is AI. As they say, "There is nothing new under the sun." Decades ago, with the advent of the internet, we were excited to add an "i" in front of product names or solutions seeking to attach our efforts to the latest and greatest tech. Now, with the ubiquity of the internet, the power of that little "i" is fleeting. This is an excellent example of the adoption of innovation as it relates to our popular naming conventions.

There are five steps to the adoption of innovation in education:

1. First we discover the innovation and try it within our existing learning efforts.

2. If we see value in enhancing our practice we use it more regularly and with more users.

3. When we have decided the innovation is worthy of widespread use, we integrate the innovation in order to learn *with* it.

4. To achieve full adoption, we then establish best practices and coach/train others to leverage a new best practice where the innovation is integral.

5. Full adoption is achieved when we stop naming the innovation as an integrated tool and it becomes a critical part of our teaching routine.

Here we have effectively revisited the Fusion Model. We must also pause and reiterate that an innovation can include, but is not solely equated to, technology. An innovation can be a shift in practice as well.[115]

Brain-Computer Interfaces

Brain-computer interfaces represent an area of technology that could potentially revolutionize education in the future but the cost is something society will no doubt need to deal with:[116]

In short, Brain-Computer Interfaces (BCIs) create direct communication pathways between the brain and an external device, typically a computer. The goal is to allow direct brain control of devices or to read brain activity for various applications. The technology behind BCIs comes in multiple forms including, non-invasive ElectroEncephaloGraphy (EEG)-based interfaces which do not require surgery or implants. They work from outside the skull and measure electrical activity in the brain. These interfaces typically use a headset or cap with electrodes that detect brain waves, a method of recording electrical activity in the brain.

Imagine a headset designed to interpret brain signals and translate them into computer commands. These wearable devices allow users to control software or devices with their thoughts including in places like virtual environments. The implications for people with impairments are profound. Imagine being able to walk despite a diagnosis of paralysis, speaking while physically unable to do so, the possibilities are endless!

The applications in education may go beyond equalizing access and abilities to potentially work to maintain student concentration by detecting when attention is wavering by monitoring brain activity. These systems could adapt learning materials in real-time based on a student's cognitive state. The power of this neurofeedback could lead to students learning to control their mental states more effectively, potentially improving learning outcomes. Initial use cases here will be limited to funding and societal acceptance.

Ethical concerns focus on privacy and consent issues related to "reading" someone's brain activity. As with any technology that interfaces directly with the brain, there are ongoing discussions about privacy, data security, and the potential for misuse. Once societal acceptance is in place, access to BCIs will be another future equity-gap like the computing device and/or internet access gaps we have already seen in the late 20th and early 21st century. Here we will find students aided by BCIs excelling in coursework and general knowledge retention in ways previously unseen leveraging BCIs well beyond their initial impairment level-setting uses to an artificial genius superpower. This challenge in the future of school makes the

calculator and then smartphone/smartwatch dilemmas schools face
seem like very small issues.

Space-Based Learning

As a child of the 1980s I learned to LOVE everything outer space-related. My life was immersed in the potential space had to offer from
Star Wars to Star Trek. Even my school days were infused with rocket
posters, NASA emblems, and the highly fashionable lazer effect
background for school photos! I drank Tang, because astronauts did,
I built model rockets to simulate blasting off into space, I loved the
idea of space. Yet the educational curriculum of this time was more
space-adjacent than space-based.

Space-based learning refers to educational experiences and
opportunities that are directly connected to or influenced by space
exploration and technology. Satellite-based internet can bridge
the digital divide, enabling online learning and teacher training in
remote areas. Advanced VR technology can simulate space environments for immersive learning experiences. Real-time data from
space missions can be incorporated into educational programs, and
some already allow students to design experiments for the International Space Station. Additionally, space-Earth comparative studies
can enhance understanding of Earth's systems, and astrobiology
and exoplanet studies offer interdisciplinary learning opportunities.
Finally, the educational potential of space technology applications,
direct communication with astronauts, and space tourism, as well as
the ethical and policy considerations surrounding space exploration.
Here is a more detailed look at these concepts:

1. Satellite-based Internet for Remote Education

Companies like SpaceX with their Starlink project are working to provide global internet coverage using satellite constellations. This could dramatically improve access to online education in remote or underserved areas worldwide. [117]

Satellite-based Internet for Remote Education:

- Bridging the Digital Divide: By overcoming the limitations of traditional terrestrial infrastructure, satellite internet can connect schools, students, and educators in areas where connectivity has been historically limited or non-existent.
- Online Learning and Resources: This accessibility enables students in remote locations to participate in online learning platforms, access educational resources, and engage with educators and peers from around the world.
- Teacher Training and Professional Development: Satellite internet also facilitates remote teacher training and professional development opportunities, ensuring that educators in underserved areas can stay up-to-date with the latest pedagogical practices and subject matter expertise.
- Educational Equity and Inclusion: The widespread availability of satellite internet has the potential to promote educational equity and inclusion, leveling the playing field for students regardless of their geographical location.
- Challenges and Considerations: While satellite internet offers promising solutions, challenges such as affordability, infrastructure development, and regulatory frameworks need to be addressed to ensure equitable access for all.
- Collaboration and Investment: Governments, educational

institutions, and private sector organizations need to collaborate and invest in satellite internet infrastructure and educational programs to fully realize the potential of this technology for remote education.

2. Space Simulations and Virtual Reality

Advanced VR technology could allow students to experience simulated space environments, "walk" on other planets, or interact with accurate models of celestial bodies. This immersive learning could make complex astronomical concepts more tangible and engaging. Advanced VR technology has the potential to revolutionize astronomy education by creating immersive simulations of space environments. Students could virtually "walk" on the surface of Mars, explore the rings of Saturn, or witness the birth of a star. These experiences would not only make complex astronomical concepts more tangible and engaging but also inspire a new generation of space enthusiasts. By interacting with accurate models of celestial bodies, students could gain a deeper understanding of the scale and dynamics of the universe. Furthermore, VR simulations could allow students to conduct virtual experiments and manipulate astronomical variables, fostering a more active and inquiry-based approach to learning. The possibilities are endless, and the potential impact on astronomy education is immense.[118, 119, 120, 121]

3. Real-time Data from Space Missions

Educational programs could incorporate live data from ongoing space missions, allowing students to analyze real scientific information. For example, NASA's educational initiatives often include data from Mars rovers or the International Space Station. Real-time Data from space missions could be integrated into educational programs, providing students with the opportunity to engage with and analyze live scientific information from ongoing space missions. This approach could foster a deeper understanding of scientific processes, data analysis techniques, and the nature of space exploration.[122]

- Data Sources and Integration: Educational programs could leverage data from various sources, including NASA's Mars rovers, the International Space Station, telescopes, and other space-based observatories. This data could be integrated into existing curricula or used to develop new educational resources and activities.

- Curriculum Development: Curricula could be designed to guide students through the process of accessing, interpreting, and analyzing real-time space mission data. This could involve activities such as tracking the position of spacecraft, monitoring atmospheric conditions on other planets, or analyzing images and other data collected by space-based instruments.

- Data Analysis and Visualization: Students could learn how to use data analysis and visualization tools to manipulate and interpret space mission data. This could involve creating graphs, charts, and other visual representations of data, as well as identifying trends and patterns.

- Collaboration and Communication: Educational programs could encourage collaboration and communication among students, scientists, and engineers involved in space missions. This could involve opportunities for students to interact with scientists and engineers through online platforms, video conferencing, or other means.

- Interdisciplinary Connections: Real-time data from space missions could be used to make connections between different scientific disciplines, such as astronomy, physics, chemistry, and biology. This could help students develop a more holistic understanding of the universe and our place within it.

- Inspiring Future Scientists and Engineers: By providing students with access to real-time space mission data and engaging them in the process of scientific inquiry, educational programs could inspire the next generation of scientists and engineers who will lead future space exploration endeavors.

4. Microgravity Experiments

Some educational programs already allow students to design experiments that are conducted on the International Space Station. This could expand in the future, giving more students hands-on experience with space research. Microgravity Experiments: The current landscape of educational opportunities already encompasses programs that facilitate student involvement in designing experiments for the International Space Station (ISS). This foundation can be significantly broadened in the future. By providing increased access and resources, a larger cohort of students could gain invaluable hands-on experience in the realm of space research. This could involve not only designing experiments but also analyzing data and interpreting results from microgravity environments, fostering a deeper understanding of scientific inquiry and the unique challenges and possibilities of space-based research. Furthermore, advancements in technology and the potential for increased commercial activity in space could open doors for more frequent and diverse microgravity experiment opportunities, potentially even extending beyond the ISS to other platforms and vehicles. Such developments could further democratize access to space research, inspiring and engaging a new generation of scientists and engineers.[123]

5. Space-Earth Comparative Studies

Learning about space environments can enhance understanding of Earth systems. For instance, studying Mars geology can provide insights into Earth's geological processes. Investigating celestial bodies and their environments can significantly enhance our comprehension of Earth's intricate systems. This comparative approach allows scientists to draw parallels and contrasts, leading to groundbreaking discoveries and a deeper appreciation for our planet's unique characteristics.[124]

Additionally, the geological formations and processes observed on Mars, such as canyons, volcanoes, and evidence of past water activity, can provide valuable insights into Earth's geological evolution. By studying the similarities and differences between the two planets, scientists can gain a better understanding of plate tectonics, volcanism, and the role of water in shaping planetary surfaces.

Furthermore, analyzing the atmospheric composition and dynamics of other planets can shed light on Earth's climate system and the factors that influence it. For example, studying the greenhouse effect on Venus can help us better understand the impact of greenhouse gases on Earth's temperature and the potential consequences of climate change.

Moreover, investigating the presence or absence of life on other celestial bodies can provide crucial clues about the origins of life on Earth and the conditions necessary for its existence. By searching for biosignatures and potential habitable environments on Mars, Europa, or other planets and moons, scientists can gain a deeper understanding of the potential for life beyond Earth and the factors that contribute to its emergence.

In addition to geology, atmospheric science, and astrobiology, space-Earth comparative studies can also encompass other fields, such as planetary physics, space weather, and comparative planetology. By examining the magnetic fields, radiation belts, and other physical phenomena of different planets, scientists can gain a better understanding of Earth's space environment and the potential impacts of space weather on our planet.

Overall, space-Earth comparative studies in K12 curriculum represent a powerful tool for advancing our knowledge of Earth and its place in the cosmos. By exploring the similarities and differences between our planet and other celestial bodies, we can gain a deeper appreciation for the complexities of Earth's systems and the delicate balance that sustains life as we know it.

6. Astrobiology and Exoplanet Studies

As we discover more about potential life beyond Earth, this could become a fascinating area of study, combining biology, chemistry, and astronomy. The burgeoning field of astrobiology, fueled by advancements in telescope technology and space exploration, is poised to revolutionize our understanding of life's potential beyond Earth. As we discover and characterize exoplanets—planets orbiting stars other than our Sun—we can delve into questions about their habitability, the potential for extraterrestrial life, and the conditions necessary for life to emerge and thrive. This interdisciplinary field combines elements of biology, chemistry, astronomy, and planetary science to investigate the origin, evolution, and distribution of life in the cosmos. Researchers in this field are developing innovative techniques to detect biosignatures—indicators of past or present life—in the atmospheres or surfaces of exoplanets. By studying the chemical compositions and physical properties of these distant worlds, we can identify potential candidates for hosting life and unravel the mysteries of our place in the universe. The search for extraterrestrial life not only expands our scientific knowledge but also raises profound philosophical and ethical questions about the nature of life, our place in the cosmos, and the possibility of encountering intelligent civilizations. [125]

7. Space Technology Applications

Learning about technologies developed for space missions and their Earth applications could inspire innovation in various fields. Integrating space technology applications into educational curricula can offer a multifaceted approach to inspire and engage students across various disciplines.[126] This could include:

- Case Studies and Project-Based Learning: Investigating technologies initially designed for space exploration and their subsequent adaptations for terrestrial use can provide real-world context for learning. Students could delve into case studies, exploring the development of technologies like GPS, satellite imaging, and solar cells, and their impact on fields such as communication, navigation, environmental monitoring, and renewable energy. Project-based learning could involve designing experiments or prototypes that utilize these technologies to address specific challenges.
- Guest Speakers and Field Trips: Inviting experts from space agencies, research institutions, or companies involved in

space technology to share their experiences and insights
can offer students a firsthand look at the possibilities
and challenges of space exploration and its technological
advancements. Organizing field trips to space museums,
observatories, or research facilities can further enhance students' understanding and appreciation of space technology
and its applications.

- Interdisciplinary Connections: Space technology applications can be integrated into various subjects, fostering
  interdisciplinary connections. For example, in science
  classes, students could explore the physics and engineering
  principles behind space technologies; in geography, they
  could analyze satellite imagery to study Earth's surface and
  climate; in social studies, they could examine the social,
  economic, and political implications of space exploration
  and its technological advancements.

- Innovation and Entrepreneurship: By understanding the
  process of innovation and technology transfer from space
  to Earth, students can be encouraged to develop their own
  innovative ideas and entrepreneurial ventures. This could
  involve identifying unmet needs, brainstorming solutions,
  and developing prototypes or business plans that leverage
  space technology applications.

- Global Collaboration: Space exploration is inherently
  a global endeavor, and its technological advancements
  have the potential to benefit all of humanity. Encouraging
  students to collaborate with peers from different countries
  on space-related projects can foster cross-cultural understanding and global citizenship.

8. Direct Communication with Space

Future technological advancements in communication infrastructure and spacecraft design could enable seamless, high-definition, and real-time video conferencing between classrooms and astronauts aboard the International Space Station or future space habitats on the Moon and Mars. These immersive experiences would allow students to directly interact with astronauts, ask questions about their daily life and research in space, and gain unique insights into the challenges and wonders of space exploration. Additionally, virtual reality and augmented reality technologies could further enhance these interactions by providing students with 360-degree views of the space environment and simulating the experience of being in space alongside the astronauts. Such advancements would not only inspire and engage students in STEM fields but also foster a deeper understanding and appreciation for the role of space exploration in expanding human knowledge and capabilities.[127]

9. Space Tourism Education

As space tourism develops, there might be educational programs preparing people for space travel or using the experience of space tourists for educational purposes. As space tourism becomes more accessible and commonplace, a new field of educational opportunities will emerge.[128] These may include:

- Pre-flight Training Programs: Specialized courses designed to prepare individuals for the physical and psychological demands of space travel. These programs could cover topics such as G-force tolerance, microgravity adaptation, basic spacecraft operation, and emergency procedures.

- Space Science and Research: Space tourism could provide opportunities for citizen science and research projects. Tourists could collect data, conduct experiments, or assist with ongoing research initiatives in space, contributing to our understanding of the universe and its effects on human physiology.
- Educational Tourism Experiences: Space tourism companies could offer educational packages that combine the thrill of space travel with learning opportunities. These experiences could include lectures, workshops, and interactive exhibits on space science, astronomy, and the history of space exploration.
- Virtual Reality and Simulation: Advanced virtual reality and simulation technologies could be used to create immersive educational experiences that replicate the sights, sounds, and sensations of space travel. These simulations could be used for both pre-flight training and educational purposes, allowing individuals to experience the wonders of space without leaving Earth.
- Curriculum Development for K-12 and Higher Education: The experiences and data gathered from space tourism could be incorporated into educational curricula at all levels. This could include developing new courses, lesson plans, and educational materials that leverage the excitement of space tourism to inspire and engage students in STEM fields.
- Professional Development for Educators: Workshops and training programs could be developed to help educators incorporate space tourism and space science into their

teaching. These programs could provide educators with the knowledge, skills, and resources needed to effectively utilize space tourism as an educational tool.

- International Collaboration: Space tourism could foster international collaboration in space education and research. Countries could work together to develop educational programs, share resources, and promote space science literacy on a global scale.

Overall, the development of space tourism has the potential to revolutionize space education, making it more accessible, engaging, and impactful for learners of all ages. By leveraging the opportunities presented by space tourism, we can inspire the next generation of scientists, engineers, and explorers and expand our understanding of the universe.

10. Ethics and Policy of Space Exploration

This could become an important area of study, considering the implications of human activity in space. As humanity ventures further into the cosmos, this field of study becomes increasingly vital.[129] It encompasses a wide range of ethical considerations arising from human activity in space, including:

- Resource Allocation and Property Rights: Who owns space resources? How should they be allocated and managed? These questions become crucial as we explore the potential for asteroid mining and resource extraction from other celestial bodies.
- Planetary Protection: How can we prevent contamination of other planets and moons with Earth organisms? How can

we protect Earth from potential extraterrestrial life forms? These questions are essential for preserving the integrity of celestial bodies and safeguarding life on Earth.

- Space Debris: How can we mitigate the growing problem of space debris and prevent collisions that could endanger future space missions? This involves developing policies for responsible space operations and debris removal.

- International Cooperation: How can we ensure peaceful and cooperative space exploration among nations? This involves developing international treaties and agreements that promote collaboration and prevent conflict in space.

- Human Rights and Space Colonization: What rights should humans have in space? How should space colonies be governed? These questions become pertinent as we contemplate the possibility of long-term human habitation beyond Earth.

- Weaponization of Space: How can we prevent the weaponization of space and ensure that it remains a peaceful domain for exploration and scientific discovery? This involves developing arms control measures and promoting disarmament in space.

- Environmental Impact: What are the environmental impacts of space exploration and how can we minimize them? This involves assessing the effects of rocket launches, space debris, and other space activities on the Earth's environment and developing sustainable practices for space exploration.

- Commercialization of Space: How should the commercialization of space be regulated to ensure equitable access and benefits for all of humanity? This involves developing policies that balance commercial interests with the public good.

- Extraterrestrial Life: What are the ethical implications of encountering extraterrestrial life? How should we interact with and protect potential alien life forms? These questions are fundamental for guiding our actions and policies in the event of first contact.

The ethics and policy of space exploration is a multidisciplinary field that draws on expertise from various domains, including law, philosophy, science, and engineering. As we continue to push the boundaries of space exploration, this field will play an increasingly important role in shaping the future of humanity in the cosmos.

It's worth noting that while some of these concepts are already being implemented to some degree, many are still speculative and depend on future technological and infrastructural developments. The extent to which space-based learning becomes a reality in 50 to 100 years will likely depend on the progress of space exploration and the accessibility of space-related technologies.

2. Societal Shifts

Globalized Curriculum

Curriculum design is an iterative, cultural process. Culture is defined as the shared beliefs, social forms, and material traits of a social group. There are five key tenets of culture:[130] Culture is learned, shared, based on symbols, integrated, and dynamic. Curriculum design is linked to culture by the way we define who we are and where we want to be from our collective shared experience. Shaping a globally-responsive curriculum takes purposeful, diligent effort in observing, selecting, and presenting experiences.[131]

Curriculum design requires we are cognizant of, and responsive to, the context in which we have been working. Historically, cultural forces have pressured an educational perspective that embraces newer, trending technology, specifically in regard to film, radio, television, and eventually computers.[132,133] Cultural forces often seek reinforcement through drawing connections with the latest innovations and by, chasing greatness of tech (a technological sublime of sorts) these forces find validation that they are connected to our future therefore remain relevant in our present.[134] To move forward in our continued iteration of curriculum, we must first look back through a historical perspective on curriculum to examine through-lines and outliers that can serve to inform our discussion.[135]

A Brief History of Curriculum

During the Hellenistic period, Greek education was largely esoteric—focused on the humanities and the curation of educated people. Students were taught to be "critical" in all, or almost all, branches of knowledge. The scope and sequence for instruction was based on the trivium: grammar, rhetoric, and logic, and the quadrivium: arithmetic, geometry, astronomy, and music. While the names are antique, the seven subjects were comparable to a modern liberal curriculum of languages, philosophy, mathematics, history, and science. The education of the Roman Empire consisted of a focus on politics and law, influenced by aspects of Greek education that focused on philosophy and language yet, reflecting on Roman-cultural priorities, centering on oratory (speech giving) due to its political significance.[136] Formalized medieval education yielded a culturally-specific, religious focus with a temporary reprieve from Greek influence. The Renaissance, or "Rebirth," yielded renewed interest in Greek and Roman systems that began transcending the geographic spheres of antiquity. In response, Renaissance education began a lasting shift back to curriculum guided by ancient wisdom.

Throughout the 1700s and 1800s, education continued to be the luxury of the elite. Efforts were made in various pockets of the globe to increase equity in regard to educational opportunity, yet education was elusive to many.

Contemporary education grew out of the Industrial Revolution as a form of mass production, a mass education, the product of the Industrial Age.[137] As writer and futurist Alvin Toffler stated, "The whole idea of assembling masses of students (raw material) to be produced by teachers (workers) in a centrally located school (factory) was a stroke of industrial genius."[138] In this system, content was broken apart into distinct units of study, veritable silos of learning. In the twentieth century, even universities felt the impact of the siloing of content. The "uni" in university became pointless, as universities' separate worlds ceased to talk to one another. Each college, department, etc. began possessing more and more autonomous power as government funding for research turned them into a loose confederation of disconnected mini-states, instead of a uni-fied organization devoted to the joint search for knowledge and truth.[139] Curriculum, teaching standards, and learning objectives all became uniquely tailored to each content category. The siloing of education has remained in K-12 classrooms, systematically validating the existing Industrial Aged model where work stations (courses) were separated and meaningful curriculum-driven collaboration is lacking.

Conversations about curriculum design were deeply impacted in the United States in the Cold War and with the Soviet launch of Sputnik. The Space Age became a cultural phenomenon yielding a hyper-focused approach that lauded math and science as the keys to a successful future, a technological sublime, that is, the pursuit of na-tionalistic greatness. From this thinking was born STEM education.

STEM education

A STEM education focuses on science, technology, engineering, and math with the belief that these will help prepare learners to compete in their future economy. STEM remains a curricular priority in our global education systems today.[140] Many policymakers, economists, and futurists believe a solid foundation in STEM will make students more marketable and successful, particularly in an increasingly globalized economy with the spread of products, technology, infor-mation, and jobs across national borders and cultures.

For decades, in the United States, multiple national policy docu-ments have asserted that global competitiveness is contingent on stu-dents being actively engaged in science, technology, engineering, and mathematics (STEM) at all levels of education.[141] In response to growing concerns about the future workforce, STEM was introduced in 2001 by the National Science Foundation (NSF) who stated: "A well-prepared, innovative science, technology, engineering, and mathematics (STEM) workforce is crucial to the Nation's prosperity and security.[142] Future generations of STEM professionals are a key sector of this workforce, especially in the critical scientific areas...To accelerate progress in these areas, the next generation of STEM professionals will need to master new knowledge and skills, collaborate across disciplines, and shape the future of the human-technology interface in the workplace."[143] Critics argue STEM itself is a socially constructed label developed in response to economic and global pressure.[144] Yet one would ask: Is that not part of the role of education? To respond to changing times? After all, edu-cation is changing because the world is changing.

The truth of the matter is that the actual educational meaning and practice of STEM is not clear. Do we approach all four as siloed curricular areas with distinct course objectives to be mastered? Or, do we instead break down the silos and leverage each of these areas around a central goal or project? I would argue the latter is where the most powerful learning potential is found. Rather than focus on the individual contents of science, technology, engineering, and mathematics, learners will make the most sense by leveraging them all together with a STEM-approach to solving real world problems. Perhaps a more realistic goal is to continue the mantra of over a millenia of education, what some call "the hidden curriculum,"[145] that is, to motivate learners to care about the future, an empathy-driven focus to cause positive change in our world.

Curricular Evolution

The world is rapidly changing as a result of both wanted and unwanted disruption. The COVID 19 pandemic has forced an increase in remote/distance and hybrid learning which has caused us to question our best practices for teaching and learning. While emergent technologies have afforded us the opportunity to teach and learn in new and exciting ways. Both of these disruptions will have lasting impacts on our culture and on our curriculum.

One example of changes in technology is AI. AI is an area of computing science focused on the creation of intelligent machines that work and react like humans. The product of impressive STEM-based work, AI learns through the data we generate in our real-time efforts and many AI products impact and improve our daily lives through automation.[146] But then automation means "machines will become very good at being machines...so we need to be extremely good at being humans again...to dig into individual abilities, allowing people to do their best and live out their potential," says innovator and futurist Liselotte Lyngsø.[147] Herein is the need to evolve our curricular focus, one that responds to a change in our global culture as seen through our understanding of the future of work. In 2018, former LinkedIn CEO Jeff Weiner stated, "Not surprisingly, there continues to be an imbalance with regards to software engineering. But somewhat surprisingly, interpersonal skills is where we're seeing the biggest imbalance. Communication is the No. 1 skill gap."[148]

Our planning and strategizing in education has often been a means to an ends approach rather than understanding that the ends would shape the means. Our focus must be on the world that our students will face. To plan without having a clear idea of what one is planning for is seemingly futile as setting out on a journey without a map. As Lahav said about future thinking, it "can serve the same role for the educational planners as did the geographers for the explorers of the Renaissance!"[149]

Real-World Focused Curriculum

If we set aside agenda-driven, esoteric rhetoric about curriculum, from a purely teaching and learning perspective, the field of instructional design offers many powerful principles that help us prescribe effective teaching and learning practices. Among them are the work of Dr. M. David Merrill who describes five principles of instruction.[150] Merrill explains that learning is promoted when: (1) learners are engaged in solving real-world problems, (2) existing knowledge is activated as a foundation for new knowledge, (3) new knowledge is demonstrated to the learner, (4) new knowledge is applied by the learner, and (5) new knowledge is integrated into the learner's world.

These five principles explain the power of hands-on learning where each individual learner makes real meaning of the process.

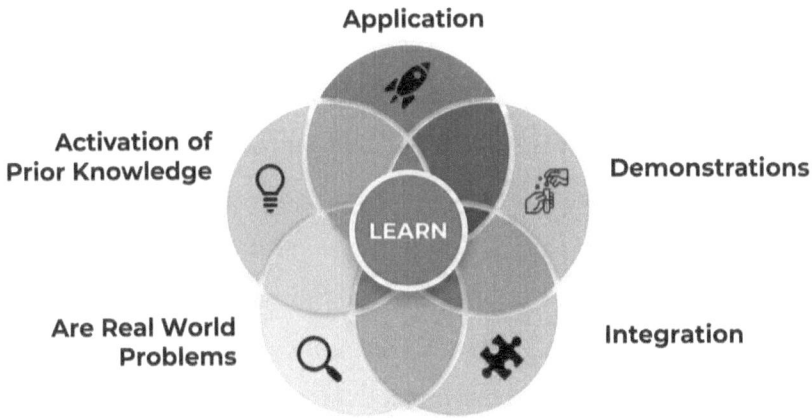

Five Principles of Instruction (Merrill, 2002).[151]

We must incorporate these principles into our curriculum design. Our future must have a problem-based approach which employs STEM-thinking, with cross-curricular content understanding, to prepare our learners for their globally conscious future. Like any innovation, curriculum change involves some overlap in practice. Changes that look and feel similar to what we know are easier to adopt. As the Fusion Model teaches us, the choice to adopt begins with an initial decision about fit as it gives way to actual implementation.[152,153,154]

STEM-thinking is a familiar set of skills that we can employ to identify important questions for important problems in real-life situations. STEM-thinking supports our explanation of both the natural and designed world through evidence-based conclusions, as a cognitive process of inquiry and an attitude that exhibits a willing-ness to engage in issues as a reflective citizen.

Meaningful adoption of a problem-based teaching with STEM-thinking serving as a lens to approach concepts can posi-tively impact learners when it is translated into policies, education programs, and practically applied in classrooms. This adoption will lead to a prepared workforce with 21st-century competencies, an advanced research agenda, and a focus on innovation. Those of us seeking to understand the future of work in an increasingly global-ized world have increased discussions about the relevance of STEM in education addressing how a STEM education leads learners to better understanding the interconnectedness of concepts and ideas.

The United Nations Sustainable Development Goals

If we put the free trade economics of globalization aside: Does STEM have potential to have a positive impact on our world? From a problem-based learning and global, cultural perspective—absolute-ly. How then do we decide, on a global level, on what problems to prepare our learners to solve?

The United Nations (UN) Sustainable Development Goals (SDGs) are 17 specific areas which serve as an urgent call for action by all countries, developed and developing, to employ global part-nerships that recognize "ending poverty and other deprivations must go hand-in-hand with strategies that improve health and education, reduce inequality, and spur economic growth, all while tackling climate change and working to preserve our oceans and forests."[155]

Sustainable Development Goals (United Nations, 2020).[156]

The SDGs provide guidance for helping cultivate focus areas for our future citizens who will understand and be ready to address them with a STEM-thinking skill set. The future we face is one together, where humankind must appreciate the global impact of our regional, geographic decisions.

A Globalized Curriculum

The global challenges we face now, and in the future, are clearly significant and will require more than an academic solution, but STEM-thinking must be part of our strategic response and be more than simply tinkering with current policies and programs, or just updating science curriculums.[157] From a humanities perspective, STEM coursework appears to tell only part of the story. Should we simply keep adding letters to the acronym like "A" for "Arts" giving us STEAM or "R" for "Reading" and/or "Research" and thus STREAM?

I would suggest that we instead accept "STEM" as a guided thought process and propose that we understand the Humanities (reading, writing, fine arts, history, social studies, etc.) as the glue that links the components of STEM together through context driven accounts, and stories, and can support STEM-thinking by emphasizing the "Why" of the story, which both explains and illustrates the meaningful application of STEM.

As an illustration of this, we must reflect on how science, technology, engineering, and math are employed to explain topics that are better illustrated by both the historical accounts of its development and the stories of modern use which illustrate how it works. Further, the humanities also serve as a civic compass to help us understand the global impact of our actions through a culturally-focused perspective, a critical component when trying to help others adopt meaningful change. An example of this is the Lucky Iron Fish story.

Lucky Iron Fish

Dr. Christopher Charles' "Lucky Iron Fish" account embodies a globalized curriculum that focuses on a real world problem that was only solved when culture was understood.

In 2008, Christopher Charles, then a masters student, traveled to Cambodia for a research project. While there, he was shocked at the high rates of iron deficiency anemia and anemia in the region. He decided to dedicate his future research to developing a safe, and affordable solution. Dr. Charles was inspired by previous research which showed that cooking in a cast iron pot increased the iron content in food. He developed an iron ingot that could be boiled in soups or drinking water. But not everyone was ready to throw a block of iron into their drinking water. It was clear that Dr. Charles had to better understand the culture in which he was working.

After doing more research on the culture, he realized what he needed to do in order to persuade people to use the ingot. Dr. Charles cast the ingot into the shape of a fish that was considered to be lucky in Cambodian folklore. As explained in his thesis, the concept of a lucky iron fish design did not pander to superstition, but created a cultural relevance for a solution based on science. To make the fish more attractive to the users, he gave the fish a smile. He called this prototype the "Happy Fish." He went on to show that almost everyone used the fish. Results from his research showed that regular use of the Happy Fish decreased anemia by 46 percent.[158]

The Lucky Iron Fish story illustrates multiple SDGs at work: #3 Good Health and Well being and #6 Clean Water and Sanitation, with a solution developed through a STEM-thinking lens that would have failed to make an impact if it was not for the humanities-directed understanding of culture.

Conclusion

Education is changing because the world is changing. Curriculum design is, and should be, in a constant state of dynamic iteration that responds to these changing times. As equitable access to education increases, so too should our desire to increase global awareness through a globalized curriculum. We have an opportunity to help plan a bright future for all humankind that transcends geography and economic status, and highlights the creative abilities we have to solve our shared problems.

Evolving workforce needs

According to the latest 2024 Gartner Hype Cycle Report: "Emotion

AI, Digital Twin of a Customer and Machine Sellers Will Have the Biggest Impact on Sales Organizations... The common theme of these three technologies are their ability to predict, interpret and serve buyers' needs and behaviors and to streamline and automate sales fulfillment..."[159] Think about these as they relate to traditional, easy-entry retail jobs. The type adults, and young adults, can fulfill due to the low barrier to entry. For young adults in particular, these are the spaces where so many life skills are learned: communication (collaboration, customer service), financial management begins, etc. Since AI will reduce the number of jobs, we now see a gap evolving in acquired skills for many. Communication and financial management will become a necessary, intentional component of the future of education.

The Future of Work

We must understand the future of education as informed by the demands from the future of work. During the pandemic people left cities where their jobs were, thinking they might be able to redefine their own future of work and be able to work from home forever. But an increasing number of employers want them back (physically) in one form or another: full time, part time, flex time, etc.[160] Longer commutes and redefined work hours/days are ripe with opportunity for mobile learning to scale.

Currently, 91 percent of internet users use mobile devices to go online at least some of the time.[161] Mobile phones are widely considered the entry point into the digital economy, and one of the most far-reaching technologies in history. [162,163]

In my post-graduate research days, I conducted a lengthy qualitative study focusing on the adoption of innovation in organizations, with the specific innovation being mobile learning (I called it "mLearning"). In this study we were creating mLearning experiences that would meet the needs of traditionally inaccessible groups.[164] To humanize the importance of the research, we (the confidential-organization and I) used the example of "Matthew" :

Matthew lives in a Tanzanian village of several thousand people. Africa's electrical grid has not found its way down the unpaved roads to Matthew's home. To access 21st-century, modern conveniences, Matthew creatively approached his environment with ideas of solutions, rather than thoughts of barriers. Matthew's refrigerator runs on a kerosene engine, he collects rainwater for drinking, and he uses solar panels for some electricity. Despite these setbacks and limitations, Matthew has mobile coverage at his home. He believes training has the potential to positively impact his work in Tanzania, but time, travel, and cost will have detrimental consequences for Matthew's work if he were to leave.

As an instructional designer (both participant and observer-researcher) on the project, I sought to transform course content, for people like Matthew, into a mobile platform allowing access to meaningful content, virtually anytime, anywhere—that was 10 years ago. The cost and computing power of smartphones has changed the landscape even more, making more powerful devices more accessible (but the accessibility gap has not closed).

In 2022, the number of smartphone users in the world was 6.6 billion, meaning 84 percent of the world's population owned a smartphone. And that is just the users, it is believed that there are over 10.4 billion mobile users worldwide, surpassing the current world population of 7.9 billion. This data means there are 2.5 billion more mobile connections than people worldwide. To be fair, not every person in the world has a mobile device, some mobile connections come from people with multiple devices, and a fraction with dual SIM's or other integrated devices like cars.[165] As the number of devices grows, so does our dependence on them to fulfill daily functions and activities.[166] In 2019, 2 billion smartphone users exclusively accessed the internet through their smartphone.[167]

On a global level, access to mobile phones is identified as a tool to bridge the learning access gap.[168,169] The experience of the on-the-go, mobile learner is increasingly recognized as important for learning traditional content, compliance training, and upskilling. The future is bright for mobile learning, the opportunities to grow limitless.

Innovation goes beyond a simple new replacement tool (technology), innovation infers enhanced application, capability, and methods of accomplishing a task—leading to a better workflow. The future of work will demand people are prepared to solve complex problems, interact with clients and colleagues alike. Further, a key indicator of our future success will be our ability to upskill in order to adapt to new tools and workflows. Over the next twenty years, we are likely to see significant disruptions to the workforce and work as we know it in recent memory. 54 percent of the US workforce says they are not confident their job would still exist in 20 years.[170] That is a scary self-reflection that must inform our instructional practice. Where is all of this change coming from?

The demographic and socioeconomic trends of the past decade, like rapid urbanization and globalization, coupled with even faster advances in technology from mobile internet to increased automation and machine learning have caused a disruption like never before.[171] Are we preparing our students to be ready for this shift in the workplace?

A trend toward a gig economy has begun. A study by Intuit predicted that by 2020, 40 percent of American workers would be independent contractors.[172] A gig based economy is defined as a way of working that is based on people having temporary jobs or doing separate pieces of work, each paid separately, rather than working for an employer.[173] Contractors developing their own brand in a feast-or-famine work atmosphere must understand collaboration, resilience, perseverance and good ol' fashioned stick-to-it-iveness! In response, our educational systems (practice, curriculum, policy, etc.) must reinforce these skills.

Asynchronous working teams

In a gig-based economy, our students will find themselves working on asynchronous teams in different geographies, different times, and with different cultures. How then can we prepare them for this workflow? Traditional, synchronous, "group work" has us creatively organize students in what we believe will be effective working teams. As a classroom teacher, I delegated these types of groups for twenty years and allowed students to pick their own teams (groups). I have also helped organize my students into teams and assign specific tasks to each person in the team. This makes a student the team "expert" in a given area on the project. But I ask myself: How often is this type of team work going to happen for my students in their future? The key point is I need to focus on their future, rather than organizing them according to my past. Asynchronous working teams are more likely what our students will be working in more often, in their future.

Clearly a technology-driven shift in education is upon us and will greatly impact the future of work for our students. This shift demands a different approach to the knowledge and skills we have relied on in the past. Our need to adopt emergent technology, and meaningfully pedagogy to leverage it, has never been this high.

Our shifts must transcend simply integrating new technology in our instruction. We must adopt innovativeness (a mindset that embraces change) in order to prepare our students for a future with a gig-based economy and its demands for new productive workflows. Helping our students understand the importance of embracing a constant state of change will help them ride out the dark valleys knowing successful mountaintops are coming.

3. Educational Access

Higher education institutions have been evolving rapidly in response to an on-demand learning need in society. This has led to increased access to higher education coursework like MIT's Open Courseware launched in 2001 and reporting in 2012 15 million users[174] grown to 210 million by 2020.[175] Further, over 30 percent of US high school students are reported as taking postsecondary college course credits in 2019.[176]

There are several key trends and implications to examine as a result of the increasing attainability (access) of college courses and degrees to high school students:

Disruption of traditional higher education models

As more students enter college with significant college credits or even associate's degrees, colleges and universities will need to adapt their offerings and value propositions. This may include shortening bachelor's degree programs from the traditional four years to two or three years. Institutions may also place a greater emphasis on specialized upper-level coursework and hands-on experiences, ensuring students are well-prepared for their chosen careers.

Furthermore, colleges and universities may choose to focus more heavily on graduate and professional programs, catering to the needs of students who have already completed a bachelor's degree. Another potential adaptation is the creation of new hybrid models that combine online and in-person learning, providing students with greater flexibility and accessibility.

Shift in the role of the traditional college experience

The future of higher education will likely see students embracing flexible learning arrangements. This includes options like commuting to campus, part-time studies while working, or fully online education. These choices can lower costs and accommodate other commitments.

Additionally, networking will expand beyond traditional campuses. Professional associations and online communities will offer platforms for building connections based on shared interests and career goals.

Furthermore, companies may play a more active role in education. Company-sponsored programs could offer specialized training tailored to specific organizational needs.

Hands-on experience will also be highly valued. Internships and apprenticeships will be crucial for applying knowledge and developing employer-valued skills.

Overall, these trends suggest that the future of education and career development will be more flexible, customized, and focused on skills development. While traditional colleges remain important, a wider range of options will emerge to meet the more immediate needs of the 21st-century workforce.

Rise of alternative credentials and portfolios

Digital portfolios are online platforms that allow individuals to showcase their work, skills, and accomplishments in an interactive format. This can include project samples, multimedia presentations, and testimonials, enabling potential employers to directly assess an individual's abilities and expertise.

Micro-credentials and badges are digital certifications that verify specific skills or competencies acquired through various learning experiences. They offer a targeted way to demonstrate knowledge in a particular area, and can be displayed on digital portfolios or social media profiles to enhance an individual's professional brand.

Competency-based assessments evaluate an individual's skills and knowledge against specific performance standards, providing a standardized way to measure competency. These assessments can be used to demonstrate mastery of specific skills, enhancing employ-ability.

Lastly, a well-crafted personal website or online presence can be a powerful tool for showcasing expertise and building a professional brand. By sharing relevant content and engaging with online communities, individuals can establish themselves as experts in their field and attract potential opportunities.

These alternative methods of demonstrating skills and knowledge offer a flexible and personalized approach to professional development, allowing individuals to adapt to the changing job market and showcase their unique talents. A clear and unique demonstration of valuable skills will set individuals apart in the future of work.

Evolution of the gig economy and hiring practices

To succeed in the ever-changing job market, future workers must take a multifaceted approach to their professional development. Cultivating versatility is essential, and can be achieved by developing a diverse range of skills applicable across various industries and job functions. This may include cross-training in different areas, pursuing continuing education, and actively seeking out experiences that broaden one's skillset.

Establishing a strong digital presence is also crucial. Building a personal brand and reputation through online platforms, such as social media, professional networking sites, and personal websites, can showcase one's expertise and attract potential employers and clients. Additionally, showcasing accomplishments through a portfolio of completed projects, testimonials, and positive client reviews provides tangible evidence of one's skills and experience.

Finally, networking and community engagement, both online and offline, are essential for building relationships, staying informed about industry trends, and accessing new opportunities. By adopting these strategies, individuals can position themselves for success in the evolving job market and remain competitive in an increasingly fluid and unpredictable professional landscape.

New models of lifelong learning and career development

The future of education will likely see a significant shift towards continuous, lifelong learning and skill development. This shift will be driven by the rapidly changing nature of work and the constant emergence of new technologies and industries, requiring individuals to continually adapt and upskill.

Subscription-based learning platforms are expected to rise in popularity, providing learners with flexible and affordable access to a vast array of courses and resources. These platforms may leverage AI-powered personalized learning recommendations to curate content and learning pathways that cater to individual needs and preferences, optimizing the learning experience for each user.

Furthermore, we can anticipate more fluid transitions between education, work, and re-skilling. This could entail closer collaboration between educational institutions and employers, and the creation of new pathways that enable learners to seamlessly navigate between different learning and work environments.

In summary, the future of education will likely be characterized by greater flexibility, personalization, and accessibility. The focus will be on empowering learners to proactively adapt and flourish in an ever-evolving world.

Looking to history, we can see parallels in how previous technological and social shifts have disrupted education and work. For example, the Industrial Revolution led to the rise of public education and standardized curricula. The Information Age brought online learning and globally distributed teams.

As we enter what some call the Fourth Industrial Revolution, marked by AI, automation, and rapid technological change, we can expect similarly profound shifts in how we prepare people for careers and demonstrate competencies.

The key for individuals will be developing adaptability, creativity, and the ability to learn continuously. For educational institutions, the challenge will be evolving to meet changing needs while preserving the core elements that make the college experience valuable.

Ultimately, I believe we will see a more diverse ecosystem of educational options and credentialing systems. Traditional degrees will remain important, but they'll be complemented by a rich tapestry of alternative pathways and ways of showcasing skills. The most successful workers will be those who can navigate this complex landscape, continually adding to their knowledge and abilities while effectively demonstrating their value to potential clients and employers.

4. Economic Factors

Automation and the changing job market

The Fourth Industrial Revolution, characterized by the fusion of digital, biological, and physical innovations, certainly has the potential to create tremendous wealth and improve living standards. However, based on historical patterns and current trends, it is unlikely to lead to universal wealth in the straightforward way some might hope.

Historically, technological revolutions have indeed generated significant wealth, but they've also tended to create or exacerbate inequalities, at least initially. The first Industrial Revolution, for instance, led to unprecedented economic growth but also to exploitation of workers and concentration of wealth among factory owners. Subsequent revolutions followed similar patterns, creating new opportunities but also new divides.

Looking at the Fourth Industrial Revolution through this lens, we can anticipate:

- Wealth creation: There will undoubtedly be enormous wealth generated through innovations in AI, robotics, biotechnology, and other emerging fields.
- Job displacement: Many traditional jobs will be automated, potentially leading to unemployment in certain sectors.
- Skills gap: The demand for high-skilled workers in

tech-related fields will likely increase, while those without relevant skills may struggle to find well-paying jobs.

- Geographic disparities: Wealth may concentrate in tech hubs and urban centers, leaving rural areas behind.
- Access inequalities: Those with access to advanced education and technology will have significant advantages over those without.

Given these factors, achieving universal wealth would require proactive measures to ensure the benefits of the Fourth Industrial Revolution are equitably distributed. This is where education becomes crucial with several trends could help address these challenges:

- Personalized learning: Adaptive platforms could help individuals acquire skills tailored to their needs and the changing job market.
- Lifelong learning: Microlearning platforms and mobile apps could facilitate continuous upskilling and reskilling throughout one's career.
- Global classrooms: Cloud-based collaboration tools could democratize access to quality education across geographic boundaries.
- Focus on soft skills: As routine tasks become automated, uniquely human skills like creativity, emotional intelligence, and complex problem-solving will become more valuable.
- Decentralized education: Blockchain-based credential verification could make education more flexible and accessible.

However, even with these advancements, achieving universal wealth would require more than just technological solutions. It would need:

- Policy interventions: Governments would need to implement policies to redistribute wealth, perhaps through universal basic income or progressive taxation.
- Corporate responsibility: Companies benefiting from automation would need to invest in retraining their workforce and local communities to which the workforce belongs.
- Global cooperation: International collaboration would be necessary to ensure that the benefits of the Fourth Industrial Revolution reach developing nations.
- Ethical considerations: Education on AI ethics and responsible use of technology will be crucial.

While the Fourth Industrial Revolution has the potential to generate unprecedented wealth, history suggests that this wealth is unlikely to be universally distributed without significant, intentional efforts. The future of education will play a pivotal role in determining how equitably the benefits of this revolution are shared. By focusing on accessible, adaptable, and lifelong learning, we can work towards a future where, if not universal wealth, at least universal opportunity becomes a realistic goal.

Reducing work hours

If we desire to realize a three to four day work week for our future we must draw direction from historical trends and current developments in education and technology.

Historically, we've seen a general trend towards reduced working hours. In the early 20th century, the standard work week was often six days long, and 10 to 12 hour days were common. Over time, labor movements and technological advancements led to the widespread adoption of the five-day, 40-hour work week in many countries. This shift was driven by increased productivity, changing social values, and a growing emphasis on work-life balance.

Looking at current trends in education and technology, several factors might contribute to a further reduction in work hours:Automation and AI: AI and machine learning are becoming increasingly prevalent in various fields, including education. This trend is likely to extend to many other industries, potentially reducing the need for human labor in certain tasks.

1. Emphasis on soft skills: These skills, such as creativity, critical thinking, and emotional intelligence, are often developed through experiences outside of traditional work environments. A shorter work week could provide more time for people to cultivate these crucial skills.

2. Lifelong learning: The concept of lifelong learning, facilitated by microlearning platforms and mobile learning apps, suggests that education will become more integrated into our daily lives. A shorter work week could allow more time for continuous learning and skill development.

3. Global collaboration: With the rise of global classrooms and advanced communication technologies, work may become more efficient and less bound by traditional time constraints.

4. Changing values: Just as the shift to a five-day work week was partly driven by changing social values, there's growing emphasis on work-life balance, mental health, and personal fulfillment that could push for further reductions in work hours.

However, it is important to note that the transition to a three to four day work week, if it occurs, is likely to be gradual and uneven across different sectors and regions. Some industries may adopt it more quickly, while others may resist the change. Economic factors, such as productivity levels and global competition, will also play a significant role in determining whether such a shift is feasible on a large scale. Two possible outcomes of a four-day work week are:

1. More travel/leisure with a boom in hospitality needs. Perhaps parents will choose a three- to four- day school week as well? How does that relate to prioritization of curriculum requirements?

2. Although a shorter work week will not necessarily lead to more leisure time or more vacations, it could lead to an increase in job stacking, that is, taking on two full time jobs. Therefore, the status quo in the school week for child care purposes of even enhancing needs to 6 to 7 days.

As demonstrated, in the above two scenarios, the impact of changes in our future work are not clear. An economic divide between groups may actually widen based on industry shifts, aptitude at leveraging new tools, and a simple mental shift in what is most valuable: time… a truly priceless commodity, may become a focus over bank accounts.

Moreover, the nature of work itself is likely to change. With increased automation and AI handling routine tasks, human work may become more focused on creative, strategic, and interpersonal roles that do not necessarily fit neatly into traditional work schedules. Specifically, we will see a growth in focus on upskilling, intuitive decision making, rational decision making, collaborative innovation, and societal systems.

1. Upskilling: Continuously Adapting to New Demands

Continuous learning and adaptability have become essential in to-day's rapidly evolving job market, where technological advancement demands constant skill development and digital literacy. Workers must demonstrate specific competencies through focused credentials while maintaining the flexibility to quickly acquire new skills as job requirements change. This emphasis on lifelong learning and digital proficiency has transformed from being advantageous to becoming a fundamental requirement for career success.

- Lifelong Learning: The rapid evolution of technology and job markets necessitates a commitment to continuous learning and skill development throughout one's life.
- Microcredentials: The acquisition of focused, industry-relevant credentials will enable individuals to demonstrate specific competencies and stay competitive.
- Skill Flexibility: The ability to learn new skills quickly and adapt to changing job requirements will be essential for career success.
- Digital Literacy: Proficiency in digital tools and technologies will be a fundamental requirement for most jobs.

2. Intuitive Decision Making: Maintaining Human Judgment to Audit AI Outputs

In an AI-driven world, critical thinking skills are becoming increasingly vital as professionals must carefully evaluate AI-generated information and identify potential biases and ethical concerns. Maintaining effective human oversight of AI systems will be crucial to ensure accountability and prevent unintended consequences.

The ability to recognize and address ethical dilemmas arising from AI-driven decisions will be a key competency for workers across industries.

- Critical Thinking: The ability to analyze and evaluate information critically, even when presented by AI systems, will be crucial.
- Ethical Considerations: Individuals must be able to identify and address ethical dilemmas that may arise from AI-driven decisions.
- Bias Detection: Recognizing and mitigating biases in AI algorithms will be essential to ensure fair and just outcomes.
- Human Oversight: Maintaining human oversight of AI systems will be necessary to prevent unintended consequences and ensure accountability.

3. Rational Decision Making: Critically Evaluating AI-Supported Decisions

The growing prevalence of AI systems demands that professionals develop strong data literacy and statistical reasoning skills to effectively evaluate AI-generated insights and make informed decisions. Workers must be able to assess both the risks and benefits of AI-supported choices while understanding their broader implications for individuals and society. This combination of data literacy, risk assessment, and systems thinking will be fundamental for navigating an increasingly AI-integrated workplace.

- Data Literacy: Understanding how data is collected, analyzed, and used by AI systems will be essential for informed decision-making.

- Statistical Reasoning: The ability to interpret statistical information and evaluate the validity of AI-generated insights will be crucial.

- Risk Assessment: Individuals must be able to assess the risks and benefits of AI-supported decisions and make informed choices.

- Systems Thinking: Understanding the broader implications of AI-driven decisions on individuals, organizations, and society will be essential.

4. Collaborative Innovation: Enhancing Human-Human and Human-AI Collaboration, Drawing from HCI Principles[177]

The future of work will increasingly depend on breaking down traditional boundaries to enable meaningful collaboration between diverse disciplines and between humans and AI systems. Successful implementation of AI technology will require a human-centered design approach that creates intuitive, trust-building interfaces that genuinely augment human capabilities. This collaborative, co-creative relationship between humans and AI will be essential for driving innovation and developing effective solutions to complex challenges.

- Interdisciplinary Collaboration: Breaking down silos and fostering collaboration between diverse disciplines will be essential for innovation.

- Human-Centered Design: Designing AI systems that are intuitive, user-friendly, and augment human capabilities will be crucial.

- Co-Creation: Encouraging collaboration between humans

and AI systems to develop innovative solutions will be essential.

- Trust Building: Establishing trust between humans and AI systems will be necessary for effective collaboration.

5. Societal Systems: Balancing AI Efficiency with Human Values and Needs in Societal Structuring

The development and deployment of AI technologies must be guided by strong principles of social responsibility and equity to ensure these powerful tools benefit society as a whole while protecting individual privacy and rights. A key challenge will be addressing bias and discrimination in AI systems while establishing democratic processes for their governance and oversight. The successful integration of AI into society will depend on striking the right balance between innovation and responsible development that prioritizes fairness, privacy protection, and inclusive outcomes for all people.

- Social Responsibility: Ensuring that AI is developed and used in a way that benefits society as a whole will be crucial.
- Equity and Inclusion: Addressing issues of bias and discrimination in AI systems will be essential to ensure equitable outcomes for all.
- Privacy Protection: Safeguarding individual privacy in the age of AI will be a major challenge.
- Democratic Governance: Establishing democratic processes for governing the development and use of AI will be essential.

It is important to maintain human qualities and critical thinking while integrating AI advancements to ensure our solutions truly meet the needs of other human beings.

While a three to four day work week is certainly possible in our future, it's not guaranteed. The path towards it will likely be influenced by technological advancements, economic factors, and societal values. As an education futurist, I believe that preparing students for this potential future will involve not just teaching them how to work efficiently, but also how to manage their time, continue learning throughout their lives, and find fulfillment both within and outside of traditional work structures.

Conclusion

The chapter examines major drivers transforming education, primarily focusing on technological revolutions and societal shifts. Within the technological sphere, it emphasizes the Fourth Industrial Revolution's impact through AI, machine learning, and brain-computer interfaces. The rise of AI is highlighted as particularly significant, with 40% of US work activities potentially being augmented or automated, necessitating a shift toward human-centric skills and AI collaboration.

A key theme is the evolution toward a globalized curriculum that moves beyond traditional STEM education. The chapter argues for breaking down subject silos and integrating humanities with STEM thinking to solve real-world problems, using the UN Sustainable Development Goals as a framework. It illustrates this approach through examples like the Lucky Iron Fish project, which succeeded by combining technical innovation with cultural understanding.

The future workforce section describes a shift toward gig economy, asynchronous teams, and reduced work hours, emphasizing five critical future skills: upskilling, intuitive decision-making, rational decision-making, collaborative innovation, and understanding societal systems. The chapter concludes by addressing educational access, noting the rise of alternative credentials, digital portfolios, and lifelong learning models as responses to these changes.

Chapter 4: The Drivers of Change - Conclusion

Drivers of Change

Forces Shaping the Future of Education

 **Technology Revolutions**

Fourth Industrial Revolution

AI and Machine Learning

Brain-Computer Interfaces

Space-Based Learning

 **Societal Shifts**

Globalized Curriculum

UN Sustainable Development Goals

Evolving Workforce Needs

Asynchronous Working Teams

Educational Access

Higher Education Disruption

Alternative Credentials

Lifelong Learning Models

Digital Divide Solutions

$ Economic Factors

Automation Impact

Universal Wealth Considerations

3-4 Day Work Week

Gig Economy Growth

 **Key Future Skills**

Upskilling   Intuitive Decision Making   Collaborative Innovation   Societal Systems

HORIZON SCAN

The forces shaping the future of education are diverse and inter-connected, from technological revolutions to societal shifts and economic factors. The Fourth Industrial Revolution, characterized by AI, brain-computer interfaces, and space-based learning, is trans-forming how we think about knowledge and skill acquisition. Mean-while, the evolution toward a globalized curriculum and changing workforce demands new approaches to teaching and learning. By understanding these drivers of change, we can better prepare our ed-ucational systems and learners for the challenges and opportunities that lie ahead.

WAYPOINTS

Key Insight 1: The Fourth Industrial Revolution is fundamentally reshaping education through AI, machine learning, and emerging technologies

Key Insight 2: Globalization and evolving workforce needs require a shift from traditional subject-based curricula to more integrated, problem-based approaches

Key Insight 3: Economic factors, including automation and the rise of the gig economy, are driving the need for new educational models and skills

Key Insight 4: Space-based learning and satellite technology present new opportunities for expanding educational access and experiences

NAVIGATION PROMPTS

🎯 Implementation

- How might your educational institution adapt to prepare students for the gig economy and remote work?
- What steps could you take to integrate global perspectives into your curriculum?

💡 Innovation

- How could AI and machine learning be thoughtfully integrated into your educational setting?
- What opportunities exist for leveraging space-based learning in your context?

🌐 Impact

- How might changes in workforce needs affect your approach to teaching and learning?
- What role could your institution play in addressing socio-economic barriers through technology?

FUTURE LOG

Observations:

Ideas to Explore:

Next Steps:

DEEP DIVES

KEY AREAS OF TRANSFORMATION

1. The Evolving Role of Educators

From knowledge transmitters to learning facilitators

In the ever-evolving landscape of education, the integration of technology and pedagogy continues to reshape the classroom experience. As we look to the future, educators, admins, and policy makers will find themselves at the forefront of a transformation that demands not only adaptability but also a profound understanding of how to leverage new tools to enhance learning.

There is a pressing need to prepare students for a world where artificial intelligence and other emerging technologies are increasingly ubiquitous. Educators are tasked with identifying how these technologies can be effectively integrated into the educational environment.

One of the most powerful methods for implementing this change is through the modeling of appropriate behavior by teachers. By demonstrating the responsible use of technology in the classroom, educators can guide students towards a mature and critical approach to these tools. For instance, a teacher might project their use of ChatGPT onto an interactive display, analyzing responses and highlighting the nuances between machine-generated content and human thought.

Exercises that challenge students to distinguish between human-created and AI-generated media become invaluable. These activities not only sharpen critical thinking skills but also foster a deeper understanding of the capabilities and limitations of AI. The transparency exhibited by educators in their own use of technology serves as a crucial model for students, demonstrating how adults navigate this new digital landscape.

The question of safety in this technological frontier is paramount. While legislation plays a role in protecting students, the most profound impact comes from the relationships formed between educators and learners. It is through these connections that values and responsible practices are truly instilled.

Looking ahead, the nature of teaching itself is poised for significant change. Two major themes emerge for the future of education: the importance of teaching students how to ask the right questions and the need for individualized learning experiences. The ability to conduct effective searches and engage in prompt engineering—skills that hearken back to the early days of Boolean logic—will be essential in a world where information is abundant but discernment is crucial.

The classroom of the future will likely focus on teaching students how to master change rather than specific technologies. This approach aligns with the problem-based learning methods already employed in some medical education programs, where the emphasis is on working backwards from real-world challenges to develop necessary skills.

Professional development for educators becomes increasingly vital. Teachers must learn to engineer experiences that cater to the individual needs of each student, much like how businesses tailor experiences for their target users. Professional development is critical to knowing that you don't know everything, a true growth mindset which is fundamental in education. There is no knowledge and skill plateau for educators.

Predicting the exact technologies that will dominate future classrooms is challenging. While virtual and augmented reality show promise, the sudden accessibility of AI tools like ChatGPT has demonstrated how quickly the landscape can shift. The key will be in maintaining flexibility and focusing on the fundamental skills that allow students to adapt to whatever technological advancements may come.

In this rapidly changing educational environment, the fusion of pedagogical expertise with technological innovation will be critical. As we navigate this transformation, the goal remains constant: to equip students with the skills and knowledge they need to thrive in an increasingly complex and technologically driven world. The journey of innovation in education is ongoing, and it is through the dedicated efforts of educators and the thoughtful implementation of new tools that we will shape the learning experiences of tomorrow.

Collaboration with AI - learning to ask the right questions

In the journey to effectively leverage the latest innovations it is critical that we pause and think back about relatable skills from our past that can inform our future. For example, the Boolean search, that is, asking the right questions to get the desired result. A formulaic approach to search has a great deal in common in what is commonly referred to as: prompt engineering.

For example, here is a Boolean search

A common example of a Boolean search method is using logical operators like AND, OR, and NOT to refine search queries. Here's a typical example:

Let's say you're searching for information about "organic farming" but want to exclude results about "industrial agriculture". A Boolean search might look like this:

"organic farming" AND farming AND NOT "industrial agriculture"

This search method would return results that:
- Include the exact phrase "organic farming"
- Contain the word "farming"
- Exclude any results that mention "industrial agriculture"

In most search engines and databases, you can use:
- AND (to narrow results, requiring all terms)
- OR (to broaden results, finding any of the specified terms)
- NOT (to exclude specific terms)

For instance, in a library database, you might search:
Agriculture AND (sustainable OR organic) NOT industrial

This approach allows users to create precise, targeted searches by combining and filtering search terms using logical operators, making it easier to find exactly the information they're looking for.

Note: The above was created by Claude.AI when I asked "what is an example of a Boolean search method?"

Prompt engineering and Boolean searches share several key principles. Both use specific terms and operators (or prompt elements) to achieve desired results, and both allow for the exclusion of certain terms or concepts. They can be combined in complex ways to achieve nuanced outcomes, and both prioritize a balance between precision and recall. Hierarchical structures and iterative refinement are essential to both processes, and they may employ controlled vocabularies or specific formats. Ultimately, both Boolean searches and prompt engineering require an understanding of the underlying system to be used effectively.

By understanding our past practice in applying these Boolean search principles to prompt engineering, we can feel more confident to create more effective, targeted, and refined prompts for AI systems. The goal in both cases is to efficiently extract the most relevant and useful information or responses from a complex system.

2. Reimagining Assessment and Credentialing

Blockchain-verified micro-credentials

Blockchain, a type of distributed ledger technology, creates a chain of data "blocks," each cryptographically linked to the previous one. It was originally developed for Bitcoin, but its potential applications extend far beyond cryptocurrencies. Key features of blockchain:[178]

- Decentralized: No single entity controls the entire network.
- Transparent: All transactions are visible to network participants.

- Immutable: Once data is recorded, it's extremely difficult to change.

In educational settings, blockchain is important for storing and verifying educational achievements. This could include: degrees and diplomas, course certificates, professional certifications, skill badges, transcripts, and training completions.[179]

Blockchain micro-credentialing represents an innovative approach to education and skill verification. The micro-credential model supports the idea of continuous, lifelong learning and allows people to gain specific skills and have them verified throughout their careers. Micro-credentials offer a more personalized, student-driven approach to education, focusing on student-centered learning enabled by technology. Micro-credentials can be earned through practical application of skills, aligning with an emphasis on experiential and hands-on learning. At the writing of this book, micro-credentialing examples include courses on LinkedIn Learning[180] and HarvardX[181] certificates, both offering tangible skill growth and validation through their institutions.

Adopting blockchain and micro-credentialing requires educators and institutions to adapt to new systems and make specific skills verifiable outside traditional degree programs, offering more accessible education. Implementing micro-credentials may require new teaching and assessment methods, fitting this book's call for pedagogical innovation.

Micro-credentials support the skill verification needed in a gig economy, which is a future trend we must continue to prepare our learners for. Blockchain verified micro-credentialing represents the kind of educational innovation that should be explored and adopted to best serve students' future needs.

One of the main motivations for using blockchain in education is to combat fraud.[182] Traditional paper credentials can be forged, and even digital records can potentially be hacked or altered. Blockchain's immutability makes it much harder to create fake credentials or tamper with existing ones. Currently, verifying someone's educational credentials often involves contacting the issuing institution directly, which can be time-consuming.[183] With blockchain, verification could potentially be done instantly and automatically. In many current systems, educational records are held by institutions, and individuals must request access or copies. A blockchain-based system could give individuals direct access to their own records, allowing them to share their credentials easily and securely with employers or other institutions.[184] If widely adopted, this technology could significantly change how we think about and handle educational credentials. It could support more flexible education models, make it easier to combine credentials from multiple sources, and potentially shift focus from institutional reputation to individual skills and achievements.[185]

It is important to note that while this technology shows promise, it's still in relatively early stages of development and adoption in the education sector. There are challenges to overcome, both technical and in terms of widespread acceptance and standardization.

Real-time, AI-powered formative assessment

The human brain is designed and ready for so much potential. We can choose a variety of amazing things in life that will lead us to powerful life experiences outside of traditional education. Each of these experiences teach us and shape us in our life pathway. I have always felt that choosing your career path at age 18 is frightening and a bit unfair. It is no surprise that we have many different jobs in our lifetime—but is that a bad thing? Think of all of the jobs you have had and how each informs some aspect of another. Now imagine being in one career your entire life as designed by a powerful AI's algorithm—remember with great power comes great responsibility.

Personalized learning is the future. How do I know this? Because personalized learning is our past. The individual, one on one, apprenticeship model allowed for responsive instructional intervention in the moment. Personalized learning is the recipe for powerful learning. Further, micro-credentialing is increasingly becoming an acceptable alternative to even a college education. Micro-credentialing is a form of certification indicating demonstrated competency/mastery in a specific skill or set of skills.[186] This can be described as a more personalized approach to learning.

The exciting vision for personalized guidance, describes, for some, a worrisome, potentially Skynet-esk solution. Personalized guidance is a definite possibility in the future and is not necessarily a bad idea, but it is going to take a lot of time and money to get right. I am a huge fan of leveraging emerging technologies to achieve more in depth learning around content and experience based objectives. I also appreciate employing these tools for increased creativity and productivity. But to leverage and employ infers, the technologies are our tools to assist in our tasks, not to direct the course of our lives. If we are uncertain of even the types of jobs that our students will have in the future, how can we place quantifiable value on current education systems? For personalization to be successful, we must operate from the premise that the world is constantly full of new jobs, needing new learning opportunities, and new instructional resources.

Tracking how people learn is not as simple as tracking choices people make. The same arguments about how many factors impact a learner's success on standardized testing—eating breakfast, a good night sleep, maturity, access to resources, etc.—can be applied to a learner's measured achievements during their whole life. Human performance technologists explore the concept of "success" for people. One of these researchers, Dr. Joe Harless, developed the "Performance Success Model," which explains the importance of motivation, environment, skills, and knowledge as contributing factors to success.[187] Are these accounted for in the algorithm? Not to mention factors like the diverse measurement tactics employed by a variety of instructors, meaning we are not comparing apples to apples for each learner across the country, let alone the world.

I am skeptical of the safety and security of public databases holding individual user account data and the implications of a breach or how that data is used by individual organizations. What if the database was hacked and the world knew about all of my detentions in middle school? If examined, quantifiably, the number of disciplinary detentions I had would have painted a terrible picture for my life's trajectory. But the qualification of these would have to include a significant amount of context, which would reveal I was a bit too social and nothing more—apparently talking out of turn is frowned upon in schools.

But I digress. The onus on schools and organizations to keep and provide meaningful and updated records is a big ask. Using this data to map potential routes through education and employment is a nice idea but I remember taking a "test" like this as a freshman in college—it directed me to being a florist or something. Of course, this product would be of higher caliber than my freshman test, but I worry about those interpreting the "potential routes" employing them as the only options available. When it comes to hard choices, sometimes we just want to be told what to do, because it is easier than navigating on our own. We would all like Godric Gryffindor's Sorting Hat[188] to make life decisions for us because it is easier. But easier does not mean better, nor does easier mean right. The *struggle* to discover our place in this world is, in fact, what leads us to that place.

Further, I am concerned that an AI-powered interface that would use this data to provide advice, may not fully understand the human experience as it relates to learning. Failure is part of our process. Our failures shape us and sometimes drive us to do more, in the same space. For example, I failed a quarter of 5th grade ELA (English Language Arts)—now I have written three books. I also failed Intro to Anthropology as a freshman in college—I have a PhD in Instructional Design, Development, and Evaluation, a field of study heavily influenced by cultural anthropology with a qualitative focus requiring extensive writing, which I have learned to love! I am not proud of my failures, but I recognize their value in pushing me forward in life.

All that being said, there is some predictability, but it is not a defined path. The intention may be for this tool to help us navigate, but when it is used to confirm/guide our path, that is when it can be dangerous. T.S. Eliot is often quoted as saying, "Most of the evil in this world is done by people with good intentions."[189] Personalized learning strategies and experiences have a great deal of potential. However, for now, personalized career pathways, as they relate to K-12, are best guided by caring individuals who choose to support the whole learner by preparing them in their life choices.

3. The Physical versus Digital Campus

Hybrid learning models and flexible educational spaces

A learning environment is the comprehensive set of physical, virtual, and psychological conditions that surround and interact with a learner, influencing their cognitive processes, attention, and ability to acquire, retain, and recall information. This encompasses tangible elements such as physical spaces, sensory stimuli, and technological interfaces, as well as intangible factors including social dynamics, digital landscapes, and the learner's internal mental state. The environment serves as the locale of cognitive presence, shaping the learner's engagement, focus, and information processing capabilities across both material and virtual realms.

The definition of environment refers to:

- The dual nature of modern learning environments (physical and virtual).
- The importance of cognitive presence in these spaces.
- The interactive relationship between the learner and their surroundings.
- The broad range of factors that can influence learning and recall.
- The idea that the environment is not just a backdrop, but an active component in the learning process.

The increasingly popular concept of flexible learning environments has emerged as a cornerstone of future education systems. Adaptive physical spaces that transcend traditional classroom boundaries, with interactive surfaces and reconfigurable furniture that transform to meet diverse learning needs, will be the norm.

Ideally, learners are transported to a world where education seamlessly blends physical and virtual realms, enabling "anytime, anywhere" learning. AI and cutting-edge technologies will facilitate personalized, flexible learning pathways, allowing students to learn at their own pace in their preferred settings, be it collaborative physical spaces, quiet individual pods, or immersive virtual reality environments.

Flexible learning environments accommodate diverse learning styles, support both collaborative and individual experiences, and adapt to the varying demands of different subjects and projects. Future educational leaders will champion and implement these adaptive spaces, using data analytics and AI to continuously optimize learning environments. By drawing parallels to current trends like flipped classrooms and blended learning models, we can uncover how today's innovations are precursors to the more advanced, adaptable environments of tomorrow. Further understanding around the Theory of Space will help us make stronger links to the past, present, and into the future.[190]

Theory of Space

Thornburg's theory of space, or "Four Spaces," is a model for describing learning spaces that can be used in schools for both students and adults:

1. Campfire: A space for learning from experts or storytellers.

2. Watering hole: A space for learning from peers.

3. Cave: A space for learning from oneself, such as through meditation.

4. Life: A space for applying what is learned to real-world projects.

Thornburg's theory is based on the idea that students learn best when they construct their own knowledge. He believes that these four learning styles have remained similar since ancient times, even in a time of constant change. He also believes that the design of learning spaces has a significant impact on the teaching and learning process.

Thornburg's theory supports the idea that learning communities should be flexible and arranged to accommodate all learning experiences and activities. He believes that flexible learning spaces, such as quiet reflection spaces, can help create psychological safety and pluralism, which are conditions that allow individuals to construct their own niches. In these niches, children can enter a "flow state" of learning, where they can collaborate, retreat, or learn. These spaces can be especially useful for autistic children, who may thrive once they've found their calling.

Thornburg's theory has also led to the development of a new learning space called the "Holodeck," which combines all four of the previous theories into one classroom space.

The physical design of learning spaces significantly impacts students' focus, collaboration, and information retention. Technology can transform teaching and learning by creating more engaging and interactive experiences, but it must be used wisely. The traditional model of education needs to be rethought, creating learning environments that are more flexible, personalized, and connected.

Research has demonstrated the profound influence of the environment on learning and knowledge retention. The physical setting plays a crucial role, with factors like lighting, temperature, and noise levels significantly impacting cognitive performance. Natural light, moderate temperatures around 70-72°F, and minimal background noise can enhance focus, attention, and memory formation. The organization of a space is equally important, as clutter can increase cognitive load and reduce working memory capacity, while well-organized environments promote better information processing.

Exposure to nature, whether through views of green spaces or the presence of indoor plants, has been shown to improve attention and reduce stress. Color psychology also plays a part, with hues like blue and green associated with increased focus and creativity. Ergonomics contribute to learning effectiveness, as comfortable seating and proper posture can minimize physical discomfort and extend attention spans. Technology can be a double-edged sword, offering enhanced learning opportunities while potentially causing cognitive strain if overused.

The social environment is another critical component of effective learning spaces. Collaborative areas that encourage peer interaction can promote knowledge sharing and reduce stress, thereby improving overall motivation. Personalization of learning spaces allows individuals to feel more engaged and comfortable, while multisensory stimulation can enhance memory formation by engaging multiple sensory inputs. Finally, air quality and ventilation are often overlooked but crucial elements that can improve cognitive function and reduce mental fatigue, ultimately supporting more effective learning and information retention.

These factors can influence attention, motivation, memory consolidation, and overall cognitive performance. Creating an optimal learning environment involves considering these elements to support effective learning and knowledge retention.

Yet, how many of these factors can be controlled in a hybrid or virtual learning environment? As we seek to inform and standardize the conditions necessary for an optimal learning environment, we must consider the rapidly evolving nature of the learning setting. The key components of an effective interactive learning environment are heavily based on the contexts by which they are employed. These contexts can be defined as:

1. Physical environment: This encompasses the overall design, layout, furniture, and resources in the learning space. It should be flexible and adaptable to support different learning activities and group sizes.[191]

2. Social environment: This involves creating opportunities for interaction between students, peers, teachers, and family members. The orientation of the space and availability of collaborative zones promote social engagement and skill-building.[192]

3. Temporal environment: This refers to how the space is utilized throughout the day, including routines, schedules, and flow of activities.[193]

4. Interactive tools and technologies: Incorporating e-learning tools, mobile apps, and virtual environments can enhance engagement and knowledge retention.[194, 195]

Additionally all learning environments benefit from intentional instructional design where the structure and presentation of content significantly impacts the learning experience. Effective instructional design should focus on promoting active participation in the form of:[196]

- Feedback mechanisms: Providing timely and constructive feedback helps learners track their progress and adjust their learning strategies.[197]
- Collaborative activities: Implementing group activities, games, and projects that encourage communication and cooperation among learners.[198]
- Adaptive evaluation: Assessing learning outcomes based on learners' cognitive characteristics rather than just behaviors.[199]
- Scaffolding: Providing appropriate support and guidance to help learners progress.[200]

Finally we cannot underestimate the importance of psychological safety, that is, creating an environment where learners feel accepted, appreciated, and motivated to engage without fear of negative consequences.[201]

An effective interactive learning environment integrates these components to create a space that is engaging, supportive, and conducive to active learning and knowledge retention. Within the world of teaching, training, and learning, the in-person realm has historically been the most tangible and relatable. The disruption of the COVID–19 pandemic led to a spike in video calls as a mechanism for training and learning. Yet post-pandemic a call back to office and in-person signaled a concern that these communication tools were simply not enough to bridge the in-person collaboration gap and from a training and learning perspective the knowledge retention ROI. Both augmented and virtual realities represent technologies primed to enhance the next level of communication technology, a place in the future of consumer life and the future of work that we must explore for the future of education.

Augmented Reality Simulations and Virtual Reality Field Trips

Augmented reality (AR) and virtual reality (VR), technologies that were once the province of science fiction and fantasy, are faster, better, and more affordable than ever. These tools have the potential to not only inspire students but to redefine how we teach and collaborate. But widespread adoption of AR and VR in K-12 classrooms requires taking risks, investing money and time, and training educators.

Innovative educators are already designing learning experiences using AR and VR that supercharge student motivation, encourage creativity, and make otherwise impossible educational adventures accessible to all.[202] Ubiquitous adoption in education has yet to be fully realized.[203] From the lens of an educator, I see some powerful educational applications for AR simulations and VR field trips where all students are immersed *together* in the learning context, and where instructors are able to measure understanding in the moment by dissecting both the content and the social interactions that occur in the learning.

To be most effective, the next steps for AR and VR in education must focus on harnessing the power of shared experience. The fictional example in *Ready Player One,*[204] a book by science fiction author Ernest Cline, illustrates this idea:

During our World History lesson that morning, Mr. Avenovich loaded up a standalone simulation so that our class could witness the discovery of King Tut's tomb by archaeologists in Egypt in AD 1922. (The day before, we'd visited the same spot in 1334 BC and had seen Tutankhamen's empire in all its glory.) In my next class, Biology, we traveled through a human heart and watched it pumping from the inside, just like in that old movie, Fantastic Voyage. In Art class we toured the Louvre while all of our avatars wore silly berets.

In my Astronomy class we visited each of Jupiter's moons.

Although it is a fictional-futurist account, Ernest Cline references an important aspect of a successful, widespread educational VR—shared experience. In each of the fictional high school courses found above, the main character attends virtual class and goes on virtual experiences with his classmates.

Virtual reality and augmented reality have come a long way over the past several years. In education we have seen applications that bring static papers to life and transport students to faraway places. Most of which is done individually, that is, the individual user on their individual device, visiting an individual place or having an individual experience.

Both VR And AR products are powerful when they embrace the concept of shared experience, a powerful way to approach learning pedagogy which leverages social interactions to promote understanding. Research around the power of shared experience in education can be understood through Albert Bandura's[205] work on self-efficacy. "Efficacy expectations determine how much effort people will expend and how long they will persist in the face of obstacles and aversive experiences."[206] Bandura further states, "The strength of people's convictions in their own effectiveness is likely to affect whether they will even try to cope with given situations."[207] To further understand efficacy expectations, Bandura used the four major sources of information as shown:

EFFICACY EXPECTATIONS

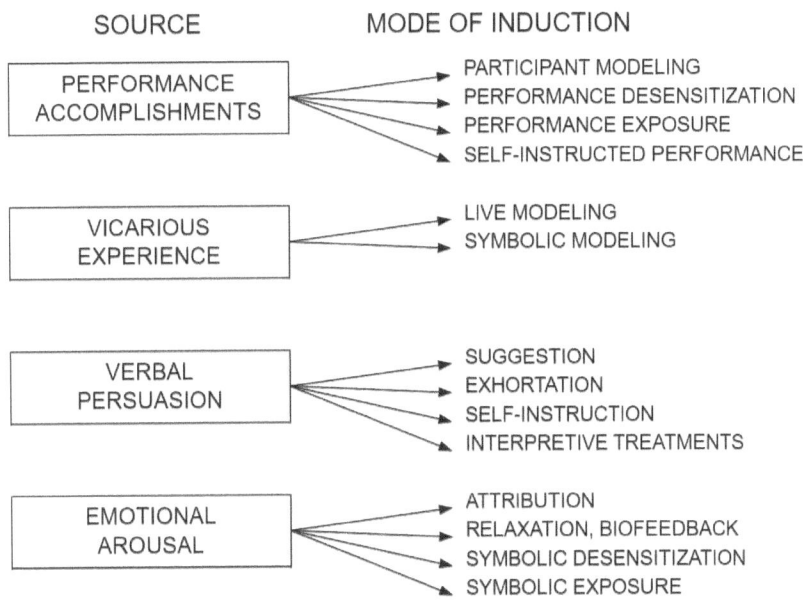

Four Major Sources Of Efficacy Expectations

Bandura's four major sources of efficacy expectations elaborate on the power of shared experience in the form of seeing others fail and social persuasion to attempt a task.[208] Seeing others succeed raises mastery expectations, likewise, seeing others perform threatening activities without adverse consequences can generate expectations in observers that they too will improve if they intensify and persist in their efforts. Finally, students can be socially persuaded that they possess the capabilities to master difficult situations by viewing other students attempt the same task.

We are a long way from Ernest Cline's description of the virtual environment (OASIS educational system[209]), and perhaps that is a good thing. I am not sure we are ready for it. Here's what Cline had to say about the educational system he developed[210]:

Question: Are the schools on Ludus your ideal institutions? If not, what is?

Cline: The virtual schools on Ludus were definitely my attempt to imagine the sort of school every nerdy kid would love to attend. A bully-free learning environment where only your brain goes to class, while your body stays at home. A school where every class-room is a holodeck, and no one ever nails you with a spitball in the back of the head. But the downside for a kid who attended a school like that would be the total lack of true human interaction and socialization. Navigating the high school maze of cliques, clubs, burnouts, and bullies helps prepare you for life after high school. In my experience, you end up using those skills a lot more than calculus or Latin.

The "lack of true human interaction and socialization" is perhaps the scary part of the future of VR in education. While seeking to leverage emergent technology to achieve educational goals, it is important that we maintain our focus and design on harnessing the power of shared-experience for every learner.[211]

Balancing screen time with physical interaction and hands-on learning

The role of physical presence plays in knowledge retention and memory recall is explained in research on spatial awareness and cognition. There is a case for an educator to be in a single location, like the front of a classroom, for directions, summary, and direct instruction. A consistent location yields higher focus and recall according to research in this space, specifically, dual-coding theory and loci.

Allan Paivio's dual coding theory suggests that verbal and non-verbal (often spatial) information are processed differently and synergistically, enhancing retention.

Additionally, the method of loci is a memory technique that involves visualizing familiar places and mentally placing information you want to remember within those locations. It's like creating a mind palace where you store information using spatial memory. By mentally revisiting these locations, you can recall the information you placed there. Research has demonstrated the effectiveness of the method of loci in improving memory performance, showing both behavioral improvements and changes in brain connectivity.

We know that incorporating spatial elements in learning materials (e.g., mind maps, diagrams) can enhance understanding and retention of complex information because spatial representations enhance learning outcomes.[212]

What does all this have to do with good teaching and learning? Interestingly, enough of the research reinforces what we have done in teaching for a really long time. A portion of instruction in the style of front-of-classroom teaching, and instructional modeling for helping students retain knowledge and skills. The case for a single location (front of classroom) for directions, summary, and direct instruction can be summarized as follows:

Dual-coding enhancement: Using a consistent front-of-classroom location leverages dual-coding theory by providing both verbal (teacher's speech) and visual (teacher's presence, gestures, and visual aids) information simultaneously. This dual-channel processing helps students encode and retain information more effectively.[213]

Spatial memory activation: A fixed instructional location activates students' spatial memory, potentially improving their ability to recall information by associating it with a specific place in the classroom. This aligns with the principles of the method of loci, where information is linked to spatial locations for better retention.

Reduced cognitive load: Consistent positioning reduces the cognitive load on students, allowing them to focus more on the content rather than constantly readjusting their attention to different locations.

As a young teacher I recall liberating myself from the front of the classroom, offering direct instruction from all over the learning space. My students would turn here and there awkwardly in their seats to try and keep up. Effectively, I had made things more difficult for them by attempting to do the opposite. Leaning more on research, I have refined by practice, and that which I promote, on more sound principles of good teaching which include:

- Enhanced visual cues: Front-of-classroom instruction allows for the integration of visual aids (e.g., whiteboards, projectors) with the teacher's explanations, providing multiple representations of information and supporting the creation of mental models.

- Improved attention management: A single focal point helps manage students' attention more effectively, reducing distractions and improving the likelihood of important information being processed and retained.

- Consistent spatial framework: Regular use of a single location for key instructional elements (directions, summaries, direct instruction) creates a consistent spatial framework. This can help students organize and recall information more effectively, similar to how mind maps and diagrams enhance understanding of complex information.

- Non-verbal communication enhancement: A fixed position allows students to more easily pick up on the teacher's non-verbal cues and gestures, which can reinforce verbal information and aid in comprehension and retention.
- Creation of a "mental stage": The front of the classroom becomes a sort of "mental stage" for students, potentially allowing them to use this space as a memory aid when recalling information later, similar to how the method of loci uses imagined spaces.

While varied instructional methods and movement can be beneficial, having a primary instructional location at the front of the classroom for key elements like directions, summaries, and direct instruction aligns well with cognitive theories of learning and memory. It provides a stable spatial and cognitive framework that can enhance students' ability to process, organize, and retain important information and skills. As we design instructional practice for the future, we need to allow the past to guide us.

4. Addressing socioeconomic barriers through technology

Moore's Law tells us technology becomes cheaper, faster, and more accessible over time. But accessibility is not the only variable in addressing adoption. The business of education has created infrastructures and processes that create barriers in and of themselves.[214] Think about the oil and gas industry, the significant financial implications of perpetuating their existence has no doubt led to their sustained business model. Think then about education: Where will we see educational changes? Perhaps in developing countries where the access needs take priority over the business aspect of education.

Historically, developing countries have leveraged emerging technologies to bridge the gap in their access to education. Let's look once again at the adoption of the radio to offer insight.

Following the pattern predicted by Moore's Law, radios would continue to become smaller, more powerful, and more user-friendly. By the end of the 1970s, 70 percent of radio receivers were either portable or mobile.[215] With better battery life and better reception, the next barrier for educational institutions was how to answer the question: How to integrate radio technology into instruction?

Instructors hoping to use radio broadcasts for instruction often had no control over content and sometimes little to no notice about what the content would be. These ground-level decision makers were hesitant to integrate technology within given educational contexts based on a lack of control over content delivery and arguably, over satisfaction with the status quo. Further, there was a constant struggle to overcome radio's one-way delivery.

In its original form, radio was a one-way communication medium, interaction with listeners was minimal. As a result, a radio program's pace was primarily that of the broadcaster (one-way, information), who found it difficult to gauge the listener's prior knowledge and attitudes, which are critical to learning.[216] To develop instructional value when using radio to replace teacher lecture, instructors began to include well-designed preparatory and follow-up materials.[217] These materials were packaged with visual and print materials and interactive elements that could be organized via listening groups.[218] The relationship between radio and instructional material became symbiotic: As materials to work with radio for instructional purposes became more effective, so too did radio broadcasts.

Interactive radio instruction (IRI) was developed in Nicaragua by a team from Stanford University in the early 1970s. The objective was to turn a typical one-way technology into a tool for active learning inside and outside of the classroom.[219] IRI efforts developed multiple techniques for using educational radio including how to function as a one-way medium used for instruction. The methodology was developed to combine the radio with the teacher to facilitate the scripted radio broadcast with a room of students through a deferred response dialogue.[220] A similar strategy was then applied by the creators of *Dora the Explorer* when Dora pauses after asking a question, hoping the viewer is responding. Over time, with increased scheduling and broadcast regulation, radio became more frequently used for instructional delivery.[221] Perhaps this will be the future of education's adoption path, those needing to increase access the most may develop new strategies that in turn inform best practices the world over.

In the near term the developing world's digital divide will be reduced by access to satellite internet. Further, new frontiers of our interstellar future on the moon, Mars, etc.. will require a reframing of curricular needs as responsively designed by the communities re-seeking to address Maslow's basic needs for their very survival. Here space becomes a new learning context. A thought exercise on what this all may look like can be inspired by a review of what leading experts are discussing at the writing of this book:

In "Toward the Stars: Technological, Ethical, and Sociopolitical Dimensions of Interstellar Exploration,"[222] author Florian Neukart examines the technological, ethical, and sociopolitical dimensions of humanity's journey toward becoming an interstellar species. He analyzes advances in propulsion systems like the magnetic fusion plasma drive, closed-loop life support systems, and habitat construction technologies that could enable long-term space travel and colonization. Beyond the technical aspects, the paper explores profound ethical considerations around terraforming other worlds, preserving extraterrestrial life, and ensuring equitable resource distribution. Neukart argues that successful space exploration requires a framework that balances scientific achievement with moral responsibility, environmental stewardship, and social justice. Further, space colonization could reshape human culture, governance systems, and evolution itself, potentially leading to distinct human subspecies adapted to different extraterrestrial environments. The conclusion is that while the technical challenges of interstellar travel are significant, the ethical and sociopolitical dimensions may prove even more critical in determining humanity's success in exploring and settling the cosmos.

With an understanding of Neukart's perspective of our potential interstellar future we can start to create a narrative sketch of what a successful citizen would look like in this future: The successful citizen in our interstellar future will need to master a complex blend of technical and humanistic capabilities. At the technical level, these individuals must demonstrate fluency with advanced technologies, particularly artificial intelligence and sustainable systems. They need to show not just the ability to use these technologies, but also to evaluate them critically and adapt as they evolve. This technical literacy extends beyond mere operational knowledge to include a deeper understanding of how these systems impact society and the environment.

Our interstellar citizens must demonstrate strong environmental stewardship and maintain deep respect for diverse life forms and ecosystems. They need to understand the moral implications of their actions and decisions, particularly as humanity extends its reach beyond Earth. This includes a profound sense of responsibility toward future generations and a commitment to ethical decision-making that considers long-term consequences.

Social abilities form another essential component of successful citizenship, here our citizens understand the importance of cross-cultural communication skills and the ability to collaborate effectively in diverse teams. Citizens understand complex social systems and participate meaningfully in democratic processes. This requires not just tolerance of different perspectives but active engagement with diverse cultures and viewpoints.

Cognitive capabilities represent a fourth crucial area. Future citizens need strong critical thinking and problem-solving skills, coupled with the ability to engage in systems thinking that recognizes interconnections between different elements of society and the environment. The research emphasizes the importance of both rational and intuitive decision-making, along with the capacity for innovation and creative thinking in approaching novel challenges.

Finally, personal characteristics that successful interstellar citizens must cultivate include resilience in the face of challenging environments, persistent curiosity and a commitment to lifelong learning. Citizens need to balance individual needs with collective responsibilities and maintain a deep appreciation for human diversity. These personal qualities underpin the technical, ethical, social, and cognitive capabilities that define effective citizenship in an interstellar society.

This comprehensive profile of the successful interstellar citizen thus, dynamically, informs the need for an educational approach that integrates technical training with ethical development, social skill building, cognitive enhancement, and personal growth. Success in this future society demands individuals who can navigate both the technological and human dimensions of an increasingly complex world while maintaining unwavering commitment to ethical behavior and environmental stewardship.

While we may have approached this thought exercise as a view into how a future society might inform future educational systems on other planets, I believe we can quickly see the tangible value of leveraging this type of futurist work to inform our current educational practice and curricular design.

Chapter 5: Deep Dives - Key Areas of Transformation Conclusion

Deep Dives

Key Areas of Educational Transformation

Evolving Role of Educators

From Knowledge Transmitters to Facilitators

AI Collaboration

Professional Development Evolution

Mentorship & Guidance Focus

Assessment & Credentialing

Blockchain-Verified Credentials

Real-Time AI Assessment

Competency-Based Evaluation

Portfolio-Based Recognition

Physical vs Digital Campus

Hybrid Learning Models

Flexible Learning Spaces

AR/VR Integration

Screen Time Balance

Socioeconomic Barriers

Technology Access Solutions

Digital Divide Bridging

Equitable Resource Distribution

Community Support Networks

Key Implementation Focus Areas

Professional Development

Technology Integration

Equity & Access

Learning Space Design

HORIZON SCAN

The transformation of education involves several key areas that require careful consideration and strategic planning. From the evolving role of educators to the reimagining of assessment and credentialing, each area presents unique challenges and opportunities. The balance between physical and digital learning spaces, coupled with efforts to address socioeconomic barriers, will shape how education serves diverse populations. Success in these transformations requires thoughtful integration of technology while maintaining focus on human connections and equity.

WAYPOINTS

Key Insight 1: The role of educators is evolving from knowledge transmitters to learning facilitators and AI collaborators

Key Insight 2: Blockchain and AI technologies are enabling new approaches to assessment and credentialing that better reflect real-world competencies

Key Insight 3: The future of learning spaces will likely blend physical and digital elements in flexible, adaptive ways

Key Insight 4: Technology can either exacerbate or help address socioeconomic barriers, depending on implementation

NAVIGATION PROMPTS

🎯 Implementation

- How might your institution support educators in transitioning to new roles and responsibilities?
- What steps could you take to create more flexible, adaptive learning spaces?

💡 Innovation

- How could blockchain technology be used to enhance credentialing in your context?
- What strategies might help balance screen time with physical interaction?

🌐 Impact

- How might changes in assessment and credentialing affect different student populations?
- What role could your institution play in addressing digital divide challenges?

FUTURE LOG

Observations:

Ideas to Explore:

Next Steps:

ETHICAL CONSIDERATIONS AND CHALLENGES

1. Data privacy and the commodification of learner information

As we collect more granular data on students' learning processes, cognitive patterns, and even neural activity, protecting student privacy and data will be paramount. There are serious ethical questions around who owns and controls this intimate data, how it can be used, and how to prevent misuse or breaches.

We are working with our planet's most precious resource: our children. And when we are working with our children, we have to be incredibly cautious with security and data privacy. We cannot be searching large language models that scrape the entire internet and provide data back to the entire internet, to reveal and exploit our children's learning challenges. Our learners should be challenged when they learn, they should fail forward, however we do not need this entire learning journey to be accessible, public knowledge. When we are working with students, we need to make sure that there are significant data security and policies in place, protecting the sanctity of the learning environment.

2. Cognitive enhancement and educational equality

There are several key ethical considerations and challenges around emerging technologies in education:

Access and Equity:

One of the biggest ethical challenges will be ensuring equitable access to advanced educational technologies like AI tutors, brain-computer interfaces, and cognitive enhancement tools. There's a real risk of exacerbating existing inequalities if only wealthy schools and students have access to these powerful learning aids. We'll need to be intentional about creating policies and funding models that democratize access.

Autonomy and Human Development:

At a global level, all industries, both public and private, will need to carefully consider how much reliance is placed on AI and further technological augmentation. There is a balance to strike between leveraging to enhance what we do each day while still fostering our ability to think independently, be creative, and develop meaningful solutions on our own human-power. Perhaps this decision actually transcends industries and begs the question: How much will society/humanity allow dependence on future innovations?

These considerations must continue to dynamically impact our policies and procedures in education reflected in our goals, assessments, cultural impact, and teacher-student relationships.

Defining Educational Goals: Innovation has and will continue to force us to grapple with fundamental questions about the purpose of education. Is the goal to optimize performance and knowledge acquisition? To cultivate wisdom and judgment? To develop so-cial-emotional skills? Our ethical framework for using these tools will depend on how we answer these questions.

Fairness and Assessment: As some students potentially gain significant cognitive advantages through technological enhancement, we will need to rethink how we assess learning and award credentials in a fair way. Traditional testing methods may no longer be equitable or meaningful (many argue they are not, nor have ever been, fair and equitable).

Cultural Implications: We will need to consider how these technologies might impact cultural diversity in education. Will AI-driven personalized learning respect and incorporate diverse cultural perspectives and ways of knowing? Or could it have a homogenizing effect? Diversity of thought and life experience are critical narratives needed to help shape a better future.

Teacher-Student Relationships: The role of human teachers will need to evolve. We will have to thoughtfully redesign practice to maintain meaningful human connections and mentorship alongside AI-driven instruction while sustaining a peer dialogue environment where meaning making is, at times, a collective experience.

Based on the history of educational technology, from the adoption of textbooks to computers in classrooms, these ethical challenges are not entirely new. However, the intimacy and power of emerging technologies like BCI (Brain-Computer Interface) and AI tutors make the stakes higher than ever before.

Ultimately, I believe we have an ethical imperative to harness these technologies to dramatically improve educational access and outcomes globally. But we must do so thoughtfully, with robust public dialogue, evolving policy frameworks, and a commitment to equity and human flourishing at the center. Science fiction authors like Jules Verne, Isaac Asimov, Gene Rodenberry, Frank Hubert, Enrest Cline, etc. have all painted a picture of a potential future which both excites and scares us. Yet, the work of science fiction should not be seen as a prescription for our future rather a description of potential. That is, we should not look to these narratives as an end point that we need to reverse engineer our innovations to fit. The narratives of science fiction should serve as inspiration for us to have open dialogue about how we prepare for the near and distant future.

We are at an inflection point where we can shape how these technologies develop and are deployed in education. Getting the ethical considerations right early on will be crucial for realizing the immense potential benefits while mitigating risks and negative consequences.

3. Balancing efficiency with humanistic values in education

To know where we are going, we must understand where we have been. Technological advancements have had a significant impact on human beings' efficacy toward achieving tasks or processes that the technology also achieves. That is, innovations impact how we see our own ability to achieve a task. Think of the calculator, "it" does math so fast, can we compete with it? Should we bother learning the math that it solves? Or think about AI tools that "write" for you. Should we continue to teach writing when it seems to work so well and fast? Of course! Just like when calculators came into education, we need to continue to understand the math to understand its uses. The calculator speeds things up, and with our knowledge speeds our processes to be more efficient. Similarly, we must understand good writing to give effective AI prompts and to meaningfully edit what we, at times, co-produce with AI. The introduction of new technologies has both enhanced and challenged human capabilities in various domains.

Research suggests that automation and technology can augment human skills and enhance productivity in certain tasks.[223,224] For example, the use of computer-aided design (CAD) software has increased the efficiency and precision of architectural and engineering design processes.[225]

Technologies like smartphones, digital assistants, and search engines allow humans to offload cognitive tasks, such as memory and information retrieval, to external devices.[226] Some studies have raised concerns about the potential negative impact of cognitive offloading on human decision-making and critical thinking skills.[227, 228]

Educational technologies, such as online learning platforms, virtual reality (VR), and augmented reality (AR), have the potential to enhance skill acquisition and learning experiences.[229, 230] However, the effectiveness of these technologies depends on their design and implementation, as well as the learner's motivation and engagement.[231]

Research suggests that effective human-machine collaboration, where humans and technologies work together in complementary roles, can enhance overall performance and efficiency.[232, 233] Achieving successful human-machine collaboration requires careful task allocation, training, and trust-building between humans and machines.[234]

While technological advancements have created new job opportunities, they have also led to job displacement in certain sectors, requiring workers to adapt and acquire new skills.[235, 236] There is a need for educational and training programs to address the potential skill gaps arising from technological disruptions.[237]

Conclusion

The complex interplay between technological advancements and human efficacy presents a multifaceted landscape of both opportunities and challenges. While technology undoubtedly has the potential to augment human capabilities and streamline various processes, it is crucial to remain cognizant of the potential downsides and unintended consequences that may arise.

Technological advancements can lead to increased efficiency, productivity, and accessibility, empowering individuals to achieve more in less time. Automation and artificial intelligence, for instance, can handle repetitive tasks, freeing up human workers to focus on more complex and creative endeavors. Furthermore, technology can facilitate communication and collaboration across geographical boundaries, fostering innovation and knowledge sharing.

However, the rapid pace of technological change also raises concerns about job displacement, skill obsolescence, and economic inequality. As automation takes over certain roles, workers may find themselves without the necessary skills to adapt to the evolving job market. Additionally, the digital divide may widen, leaving those without access to technology at a disadvantage.

Ethical considerations also come into play, particularly concerning issues such as privacy, data security, and algorithmic bias. The collection and use of personal data by technology companies raise questions about surveillance and control, while algorithms used in decision-making processes may perpetuate existing biases and discrimination.

Therefore, it is imperative to adopt a proactive and responsible approach to technological development and implementation. Ongoing research and critical analysis of the social, ethical, and economic implications of new technologies are essential. Policymakers, industry leaders, and researchers must collaborate to ensure that technological advancements are harnessed for the benefit of society as a whole, while minimizing potential harm and addressing unintended consequences.

By navigating the complexities of the technology-human efficacy relationship with foresight and ethical considerations, we can create a future where technology truly enhances human capabilities and fosters a more equitable and just society.

Chapter 6: Ethical Considerations and Challenges - Conclusion

Ethical Considerations

Challenges in Future Education

🛡 Data Privacy

Student Data Protection

Information Ownership

Data Usage Guidelines

Security Protocols

⚖ Equity & Access

Technology Accessibility

Resource Distribution

Digital Divide Impact

Economic Barriers

♡ Human Values

Personal Connection

Social Development

Cultural Preservation

Emotional Intelligence

🧠 AI Ethics

Algorithm Bias

Decision Transparency

Human Oversight

Assessment Fairness

→ Key Principles

Privacy by Design

Universal Access

Human-Centered

Ethical AI Use

HORIZON SCAN

As we embrace new technologies and approaches in education, ethical considerations become increasingly critical. The challenges of data privacy, cognitive enhancement, and maintaining humanistic values require careful navigation. We must ensure that our pursuit of innovation and efficiency doesn't compromise student privacy, exacerbate inequalities, or diminish the human elements that make education transformative. Success in future education depends not just on technological advancement, but on our ability to implement these advances ethically and equitably.

WAYPOINTS

Key Insight 1: Data privacy and the responsible use of student information must be prioritized as we collect more detailed learning data

Key Insight 2: The ethical implications of cognitive enhancement technologies require careful consideration and clear guidelines

Key Insight 3: Maintaining human connection and values in education becomes more crucial as technology plays a larger role

Key Insight 4: Balancing efficiency with equity and accessibility remains a central challenge in educational innovation

NAVIGATION PROMPTS

🎯 Implementation

- How might your institution better protect student privacy while leveraging learning analytics?
- What guidelines could help ensure ethical use of AI and other emerging technologies?

💡 Innovation

- How can you maintain focus on human connection while implementing new technologies?
- What strategies could help balance technological efficiency with humanistic values?

🌍 Impact

- How might different ethical frameworks affect technology adoption in your context?
- What steps could ensure that innovation promotes rather than compromises educational equity?

FUTURE LOG

Observations:

Ideas to Explore:

Next Steps:

PREPARING FOR THE FUTURE

RECOMMENDATIONS AND STRATEGIES

2059 will come faster than we might expect. In preparation, each of us needs to understand our place and potential to positively affect change in education. Our strategic approach must focus on the need to adapt to technological advancements and respond to societal changes. Our emphasis must be a human-centric approach that prioritizes competency-based progression, lifelong learning, and globalized, project-based curricula. Accessibility and inclusivity must remain a priority in education, as well as a shift in the role of educators from knowledge transmitters to learning facilitators. Additionally, ethical considerations such as data privacy and the balance between technology and the humanized learning experience must be at the forefront of planned change. Intentional investment, collaboration, and a focus on return on instruction will help us to create a future-ready education system that empowers all learners.

1. Policymakers

We stand at the cusp of unprecedented technological and societal change, it is imperative that we reimagine our education system to prepare learners for a world vastly different from our own. By examining historical patterns and leveraging emerging innovations, we can create an education framework that is both visionary and pragmatic.

The Evolution of Educational Paradigms

Throughout history, education has evolved in response to societal needs. From the Hellenistic focus on humanities to the Industrial Revolution's standardized mass education, each era has shaped learning to meet its demands. As we enter the Fourth Industrial Revolution, characterized by the fusion of digital, biological, and physical innovations, we must once again adapt our educational approach.

Technological Integration and Human-Centric Learning

While artificial intelligence, virtual reality, and brain-computer interfaces offer exciting possibilities, we must remember that technology should amplify, not replace, human elements in education. As the U.S. Department of Education aptly stated, we should envision "a technology-enhanced future more like an electric bike and less like robot vacuums." Our goal should be to create learning environments where technology enhances human potential.

Competency-Based Progression and Lifelong Learning

The traditional age-based grade system is becoming obsolete. By 2059, I envision a shift towards competency-based progression, allowing learners to advance at their own pace. This aligns with the growing need for lifelong learning in a rapidly changing job market. Policies should support flexible learning pathways and recognize diverse forms of credentials, including micro-credentials verified by blockchain technology.

Globalized Curriculum and Project-Based Learning

Our curriculum must evolve to address global challenges. The United Nations Sustainable Development Goals provide an excellent framework for developing a globalized curriculum that prepares students to tackle real-world issues. Project-based learning, integrating STEM thinking with humanities, will be crucial in fostering critical thinking and problem-solving skills.

Accessible and Inclusive Education

As technology becomes more pervasive, we must ensure equitable access to educational resources. Historically, developing countries have leveraged emerging technologies to bridge educational gaps. We should learn from these examples and invest in infrastructure that supports universal access to high-quality education, regardless of geographical or socioeconomic barriers.

The Evolving Role of Educators

Teachers will transition from knowledge transmitters to learning facilitators. Continuous professional development in emerging technologies and pedagogies will be crucial. Policies should support this transition, providing resources for teacher training and recognizing new forms of educational leadership.

Data Privacy and Ethical Considerations

As we collect more granular data on student learning, robust policies must be in place to protect student privacy and ensure ethical use of this information. We must strike a balance between leveraging data for personalized learning and safeguarding individual rights.

Funding and Resource Allocation

To realize this vision, we need strategic investment in educational infrastructure, teacher development, and research. The Fusion Model for organizational adoption of innovation provides a framework for effectively implementing and scaling educational innovations. By understanding the stages of innovation adoption—from agenda setting to routinizing—we can allocate resources more efficiently and effectively.

Conclusion

As policymakers, you have the power to shape the future of education. By embracing a culture of innovation, investing in flexible learning environments, and prioritizing equity and accessibility, we can create an education system that truly prepares our learners for the challenges and opportunities of 2059 and beyond.

Remember, as history has shown us time and again, education is the cornerstone of societal progress. Let us work together to build an educational framework that empowers every learner to reach their full potential and contribute meaningfully to our shared future.

2. School Administrators

School Administrators both in central offices and in school buildings, have a practical focus that is not merely on return on investment, but more critically, on return on instruction. By examining historical patterns and leveraging emerging innovations, administrators can create an education framework that is both visionary and pragmatic.

The Imperative of Strategic Innovation Adoption

Throughout educational history, from the integration of textbooks to computers, the adoption of new technologies has been a constant challenge. As we navigate the Fourth Industrial Revolution, characterized by the convergence of digital, biological, and physical innovations, the stakes for effective technology integration are higher than ever. The Fusion Model provides a robust framework for understanding and implementing this process.

Leveraging the Fusion Model for EdTech Implementation

To effectively lead an EdTech project, employ the Fusion Model's stages:

A. Agenda Setting: Define clear, SMART goals aligned with educational objectives. It's crucial to set Specific, Measurable, Achievable, Relevant, and Time-bound (SMART) goals that resonate with both the educational mission and the technological possibilities.

B. Matching: Involve stakeholders to select appropriate innovations.

C. Redefining/Restructuring: Provide training and adjust organizational structures.

D. Clarifying: Maintain open communication and conduct risk assessments.

E. Routinizing: Establish evaluation metrics to track progress towards full adoption.

This approach fosters a culture of innovativeness, driving meaningful change in education.

Measuring Return on Instruction

While traditional Return on Investment (ROI) is important, your focus must be on Return on Instruction. This requires a multifaceted approach:

- Student Outcomes: Track improvements in academic performance, engagement, and skill development.
- Teacher Effectiveness: Measure how technology enhances pedagogical practices and reduces administrative burdens.
- Personalized Learning: Evaluate the extent to which technology enables individualized instruction and adaptive learning paths.
- Data-Driven Decision Making: Assess how data analytics inform instructional strategies and interventions.

Much of which is directly related to current practice, this is an example of how we need to keep revisiting our current strategies to reflect on their meaningfulness as time goes on.

Pilot Programs and Iterative Implementation

Small scale pilot programs are effective in allowing to us test new educational technologies on a small scale before broader implementation. The Fusion Model provides a comprehensive framework for evaluating new tools beyond mere usability, considering how they fit within your school's culture, systems, and goals. This approach helps assess user engagement, measure learning outcomes, identify technical issues, evaluate cultural fit, consider long-term sustainability, and analyze integration with existing systems.

Professional Development and Change Management

Invest in intentional, ongoing professional development for educators. The success of any EdTech initiative hinges on teachers' ability to effectively integrate technology into their instructional practices. Create a supportive environment where educators feel empowered to experiment with new tools and pedagogies. Organizational adoption of innovation comes to full fruition (Routinizing) when individuals can articulate the value of the innovation for themselves and those around them. Professional Development can be a perfect platform for these conversations.

Data Privacy and Security

As you collect more granular data on student learning, robust policies must be in place to protect student privacy and ensure ethical use of this information. Balance the potential of data-driven personalized learning with the imperative to safeguard individual rights.

Flexibility and Adaptability

Maintain a nimble approach, allowing for quick pivots when faced with unexpected outcomes. This might involve switching tools, adjusting teaching methods, or revising objectives. By maintaining a flexible mindset, you can swiftly address challenges and create innovative solutions that align with your goals. Here is a place for us to exemplify a fundamental trait for both ourselves and our learners: agency. Discussed more in the next chapter, this is the stick-to-it-tiveness and general rigor each of us must employ as we embark in a world of constant change.

Collaboration and Knowledge Sharing

Expand beyond your personal network when addressing educational technology challenges. Embrace diversity of thought by engaging with individuals from different backgrounds, disciplines, or even industries. This approach fosters cross-pollination of ideas, leading to more robust, innovative, and inclusive solutions for educational technology challenges.

Long-Term Vision and Sustainability

While immediate results are important, focus on long-term impact. Consider how your technology investments align with future trends in education and workforce needs. The goal is not just to improve current practices, but to prepare students for a rapidly evolving world.

Conclusion

As school administrators, you are at the forefront of shaping the future of education. By leveraging the Fusion Model and focusing on return on instruction, you can make informed decisions about technology investments that truly enhance learning outcomes. Remember, the goal is not to adopt technology for its own sake, but to create learning environments that empower students and teachers alike.

Your role is crucial in bridging the gap between visionary educational goals and practical implementation. By carefully validating your technology investments with verifiable results, you ensure that your school not only keeps pace with technological advancements but leverages them to create truly transformative learning experiences.

The journey of innovation in education is ongoing, and it requires your leadership, vision, and commitment to continuous improvement. By focusing on return on instruction, you are not just managing resources; you are shaping the future of learning itself.

3. Educators

As a futurist specializing in education, with experience as both a teacher and consultant to education leadership, I present these guidelines for educators tasked with preparing students for an uncertain yet exciting future. Our mission is to equip learners with the skills, mindset, and adaptability needed to thrive in a rapidly evolving world. By examining historical patterns and leveraging emerging innovations, we can create an educational approach that is both visionary and pragmatic.

Embracing the Evolution of Pedagogy

Education has continually transformed alongside human civilization. Ancient Greece gave us dialectical learning through Socrates' questioning method, while the Industrial Revolution introduced standardized classroom instruction to prepare workers for factory life. Now, as we enter an age where digital technology, biological advances, and physical systems increasingly merge, our educational approaches must undergo another fundamental shift to meet these new realities.

Embracing lifelong learning and adaptability

- Cultivate an innovator's mindset—be willing to try new things, take risks, and learn from failure.
- Continuously update skills and knowledge, especially related to emerging technologies and pedagogies.

- Design learning experiences that build students' adaptability, creativity and problem-solving skills.
- Leverage technology to personalize learning and provide students with more agency.
- Collaborate with colleagues to share ideas and create a culture of innovation in schools.

To create effective instructional materials, begin by analyzing your audience, context, and content. Consider learner characteristics, learning environment, subject complexity, and available resources. Select appropriate media types including text, audio, video, images, animations, and interactive elements. Choose suitable formats such as print, digital, online, or blended. Employ a mix of synchronous and asynchronous delivery methods to accommodate diverse learning needs. Ensure chosen methods and formats align with educational objectives and technological capabilities. This comprehensive approach facilitates successful implementation and integration of instructional materials within the educational organization, supporting the overall adoption of innovative teaching practices.

Cultivating an Innovator's Mindset

The Fusion Model provides a framework for understanding and adopting innovation in education. As educators, we must first cultivate the mindset of an innovator, the one who tries new tools and openly reflects on the tools' meaningful, or meaningless, role in amplifying good practice. This involves:

A. Agenda Setting: Identify areas in your teaching practice that need innovation.

B. Matching: Explore new pedagogical approaches and technologies that align with your goals.

C. Redefining/Restructuring: Adapt these innovations to fit your unique classroom context.

D. Clarifying: Implement new practices, reflecting on their effectiveness.

E. Routinizing: Integrate successful innovations into your regular teaching practice.

Think of the first 1:1 device teachers who would blog about going paperless and talk about what they learned. These brave, transparent educators helped many of us to see we could do it too! We need more such teachers as we embrace the future of learning. By modeling this process, we provide students with a tangible example of lifelong learning and adaptability.

Focusing on Future-Ready Skills

While content knowledge remains important, our primary focus should shift to developing skills that will remain relevant in an AI-driven world:

- Critical Thinking and Problem-Solving: Teach students to analyze information, identify patterns, and develop creative solutions.
- Adaptability and Lifelong Learning: Foster a growth mindset and the ability to continually acquire new skills.
- Digital Literacy: Ensure students can navigate, evaluate, and create digital content responsibly.
- Emotional Intelligence: Cultivate empathy, self-awareness, and interpersonal skills.

- Collaboration and Communication: Prepare students to work effectively in diverse, often virtual, teams.

Leveraging Technology as a Learning Amplifier

Technology should enhance, not replace, human-centered learning. Use tools like artificial intelligence, virtual reality, and adaptive learning platforms to:

- Personalize learning experiences.
- Provide immediate feedback.
- Create immersive, experiential learning opportunities.
- Connect students with global peers and experts.

Remember, the goal is to use technology to amplify good teaching practices, not to rely on it as a substitute for effective pedagogy. With that in mind, be ready to give yourself permission to pause, or slow down, integration of new tools until you believe the positives outweigh the barriers.

Fostering Agency and Self-Directed Learning

Change is hard. Students need to see adults managing change and be vicariously guided in order to prepare for their future where they will need to continuously upskill and reskill. Encourage agency by:

- Offering choices in learning paths and projects.
- Teaching self-reflection on the learning process (metacognitive strategies).
- Encouraging self-assessment and reflection.
- Providing opportunities for student-led inquiry.

Creating Authentic, Real-World Learning Experiences

Move beyond siloed subjects to interdisciplinary, project-based learning that reflects real-world challenges. Consider aligning projects with the United Nations Sustainable Development Goals to give students a global perspective and sense of purpose. Revisit the Lucky Iron Fish story for inspiration!

Embracing Failure as a Learning Opportunity

In a rapidly changing world, the ability to learn from failure is crucial. Create a classroom culture that views mistakes as opportunities for growth. Model this by being transparent about your own learning process and challenges in adopting new practices.

Continuous Professional Development

Stay informed about emerging trends and technologies that may impact education. Engage in professional learning networks, attend conferences, and participate in online courses. Share your learning journey with your students, modeling lifelong learning.

Balancing Technology with Human Connection

While technology offers powerful learning tools, remember the irreplaceable value of human connection in education. Prioritize building strong relationships with your students and fostering a supportive classroom community.

Preparing for an AI-Augmented Future

Artificial intelligence is increasingly becoming part of our daily lives. Teach students to work alongside AI tools as previously discussed in the section on "AI and the rise of editors" in Chapter 4.

- Develop prompt engineering skills.
- Critically evaluate AI-generated content.
- Understand the ethical implications of AI.
- Identify tasks best suited for human vs. AI capabilities.

Conclusion

As educators, we have the profound responsibility and privilege of shaping the future through our students. By adopting an innovator's mindset, focusing on future-ready skills, and leveraging technology thoughtfully, we can prepare our students not just for the jobs of tomorrow, but for a lifetime of learning and adaptation.

Remember, the most powerful tool in your classroom is not any piece of technology, but your ability to inspire, guide, and empower your students. Your role is evolving from a transmitter of knowledge to a facilitator of learning experiences. By providing students with access to your own agency—your process of figuring things out and adapting to change—you equip them with the most valuable skill for their future: the ability to learn, unlearn, and relearn throughout their lives.

The future of education is not about predicting exactly what knowledge or skills will be needed, but about fostering the curiosity, resilience, and adaptability that will allow our students to thrive in any future they encounter. Through your dedication and innovation, you are not just teaching subjects; you are preparing the problem-solvers, innovators, and leaders of tomorrow.

4. Students

The journey ahead of you is filled with incredible opportunities, and preparing for your future is about more than just grades and test scores. By developing your sense of agency and embracing lifelong learning, you will build the foundation for success in a rapidly changing world. You have the power to shape your path by mastering essential skills like critical thinking, digital literacy, and AI fluency, while also nurturing your emotional intelligence and global awareness. Do not be afraid to take calculated risks, start entrepreneurial ventures, or dive into interdisciplinary projects that combine your various interests. Remember that failure is not a setback but a stepping stone to growth, and maintaining your mental and physical well-being is just as important as academic achievement.

The future workplace will value those who can adapt, innovate, and collaborate across cultures and disciplines, so seek out opportunities to expand your horizons through online learning, cultural exchanges, and meaningful projects. By developing a strong ethical framework and building your personal brand while staying committed to sustainability, you will be well-equipped to face the challenges and seize the opportunities that lie ahead. Your generation has the unique advantage of growing up in a digital age, and by embracing these skills and mindsets now, you're not just preparing for the future—you are preparing to shape it.

Develop Your Sense of Agency

Agency is your capacity to act independently and make your own choices. It is a critical skill for navigating an uncertain future. To cultivate agency:

- Take ownership of your learning journey. Set personal goals beyond what is assigned in class.
- Seek out opportunities to make decisions and solve problems independently.

- Reflect on your choices and their outcomes to build decision-making skills.

Embrace Lifelong Learning

The rapid pace of technological change means that learning cannot stop when you leave school. Adopt a growth mindset and commit to continuous learning:

- Explore online learning platforms like Coursera, edX, or Khan Academy to supplement your formal education.
- Develop the habit of reading widely across various subjects.
- Pursue passion projects and hobbies that challenge you to acquire new skills.

Master the Art of Learning

Learning how to learn effectively is perhaps the most valuable skill you can develop. Focus on:

- Understanding your learning preferences and strategies that uniquely help you to succeed.
- Developing strong study habits and time management skills.
- Practicing metacognition, which is thinking about your thinking and learning processes.

Build Digital Literacy and AI Fluency

As we enter the Fourth Industrial Revolution, digital skills are non-negotiable:

- Familiarize yourself with a range of digital tools and platforms.

- Learn the basics of coding and data analysis focusing on what questions to ask (prompts) more than how to ask them (coding).
- Develop prompt engineering skills for AI tools like ChatGPT.
- Practice critical evaluation of online information and AI-generated content.

Cultivate Future-Ready Skills

While specific job requirements may change, certain skills will remain valuable:

- Critical thinking and problem-solving
- Creativity and innovation
- Communication and collaboration
- Emotional intelligence and empathy
- Adaptability and resilience

Seek out opportunities to develop these skills both in and out of the classroom by writing blogs, creating videos, publishing music, and recording podcasts.

Engage in Project-Based and Interdisciplinary Learning

The future workplace will require the ability to integrate knowledge from various fields:

- Participate in extracurricular activities that combine multiple disciplines.
- Propose interdisciplinary projects to your teachers, that is, look for links to explore between science and history, or english and art (as examples).

- Apply what you learn in one subject to challenges in another.

Gain Global Perspective

In an increasingly interconnected world, global awareness is crucial:

- Learn a second (or third) language.
- Engage with international news and global issues.
- Participate in cultural exchange programs or virtual collaborations with students from other countries.

Develop Entrepreneurial Skills

The gig economy and rapid job market changes make entrepreneurial thinking valuable:

- Start a small business or side project.
- Participate in entrepreneurship clubs or competitions.
- Practice identifying problems and developing innovative solutions.

Focus on Sustainability

Climate change and environmental issues will be significant challenges for your generation:

- Educate yourself about environmental issues and sustainable practices.
- Participate in environmental initiatives in your school or community.
- Consider how sustainability relates to your career interests.

Build a Personal Brand and Network

Your online presence and professional network will be increasingly important:

- Develop a professional online presence (e.g., LinkedIn profile).
- Attend industry events or webinars in fields that interest you.
- Seek out mentorship opportunities.

Practice Ethical Decision-Making

As technology advances, ethical considerations become more complex:

- Study ethics and philosophy to develop a strong moral framework.
- Discuss ethical dilemmas related to technology and AI with peers and mentors.
- Consider the ethical implications of your career choices and personal actions.

See Chapter 4's discussion of Ethical Decision Making Societal Systems: Balancing AI Efficiency with Human Values and Needs in Societal Structuring, which discusses key areas of concern in AI development and implementation: ensuring AI benefits society, addressing bias and discrimination in AI systems, protecting individual privacy, and establishing democratic governance for AI.

Prioritize Mental and Physical Health

Maintaining well-being will be crucial in a fast-paced future:

- Develop stress management techniques.
- Prioritize physical exercise and healthy habits.
- Practice mindfulness or meditation to enhance focus and emotional regulation.

Embrace Failure as a Learning Opportunity

In a rapidly changing world, the ability to learn from failure is crucial:

- Take calculated risks and step out of your comfort zone.
- Reflect on failures to extract valuable lessons.
- Develop resilience by persevering through challenges.

Conclusion

Remember, the goal is not to predict the exact skills or knowledge you will need in the future, but to develop the adaptability, curiosity, and resilience to thrive in any scenario. By focusing on these areas, you are not just preparing for a specific job or career path, but for a lifetime of learning, growth, and meaningful contribution to society.

Your education is not confined to the classroom. Every experience is an opportunity to learn and grow. Embrace the uncertainty of the future as an exciting challenge, and trust in your ability to adapt and evolve. With a strong foundation in these areas, you will be well-prepared to navigate the complexities of the future and create positive change in the world.

5. Parents

As parents, you hold the incredible power to shape your children's future by fostering the skills and mindsets they will need to thrive in an ever-changing world. By nurturing your children's sense of agency and independence through age-appropriate decision-making, while modeling your own problem-solving approaches and learning journey, you are laying the groundwork for their future success. In today's digital age, balancing technology literacy with unplugged experiences, critical thinking, and creative exploration is crucial. Remember that preparing children for tomorrow goes beyond academic achievement—it is about cultivating emotional intelligence, environmental consciousness, and global awareness through hands-on experiences and meaningful family discussions.

By encouraging a growth mindset where failures become learning opportunities, supporting diverse interests from arts to entrepreneurship, and promoting financial literacy alongside soft skills like communication and teamwork, you are equipping your children with the adaptability and resilience they'll need. While we cannot predict exactly what the future holds, by creating an environment rich in curiosity, creativity, and continuous learning, you are not just preparing your children for specific scenarios—you are empowering them to confidently navigate any challenges and opportunities that come their way. Your investment in their holistic development today, balanced with plenty of unplugged time for physical activity and face-to-face interactions, will help them become well-rounded individuals ready to contribute meaningfully to tomorrow's world.

Foster Agency and Independence

Agency—the capacity to act independently and make free choices—is crucial for future success. To cultivate this in your children, allow them to make age-appropriate decisions and face the consequences. Encourage problem-solving rather than immediately providing solutions. Assign responsibilities at home to develop a sense of capability and independence.

Embrace the Power of Access to Agency

Children learn significantly through observation and imitation. Provide access to your own agency by demonstrating how you approach challenges and learn new skills. Share your thought processes when making decisions or solving problems. Be transparent about your own learning journey and struggles with new technologies or concepts.

Cultivate a Growth Mindset

Encourage a belief that abilities can be developed through dedication and hard work. Praise effort and strategy rather than innate talent. Frame failures as learning opportunities. Model resilience and perseverance in your own pursuits.

Promote Digital Literacy and Responsible Technology Use

In our increasingly digital world, teach critical evaluation of online information and AI-generated content. Discuss responsible digital citizenship and online safety. Encourage creative and productive uses of technology beyond passive consumption.

Support Holistic Development

Focus on developing well-rounded individuals. Encourage participation in diverse activities like arts, sports, and community service. Foster emotional intelligence through open discussions about feelings and relationships. Promote physical health and wellness as foundational to overall success.

Encourage Curiosity and Lifelong Learning

Instill a love for learning that extends beyond formal education. Explore topics of interest together through books, documentaries, or educational outings. Engage in family learning activities, like learning a new language or skill together. Model lifelong learning by pursuing your own educational goals.

Develop Global Awareness and Cultural Competence

Prepare children for an interconnected world by exposing them to diverse cultures through food, music, literature, and travel if possible. Discuss global events and their impacts at an age-appropriate level. Encourage learning a second language from an early age.

Foster Creativity and Innovation

Creativity will be vital in an AI-driven future. Provide open-ended toys and activities that encourage imaginative play. Ask open-ended questions that prompt creative thinking. Create a home environment that values and celebrates creativity.

Teach Financial Literacy

Prepare children for a changing economic landscape by introducing basic financial concepts early. Provide opportunities to earn and manage money. Discuss various career paths and the evolving nature of work.

Emphasize Soft Skills

While technical skills are important, soft skills remain crucial. Encourage teamwork through group activities or sports. Practice active listening and effective communication at home. Teach conflict resolution skills through real-life situations.

Nurture Environmental Consciousness

Prepare children to address future environmental challenges by engaging in eco-friendly practices at home. Discuss environmental issues and potential solutions. Participate in community environmental initiatives as a family.

Encourage Entrepreneurial Thinking

Foster an entrepreneurial mindset by supporting children's initiatives to solve problems or create products. Discuss business concepts using real-world examples. Encourage participation in youth entrepreneurship programs.

Promote Hands-On, Project-Based Learning

Complement formal education with practical experiences. Engage in DIY projects together. Encourage participation in science fairs, coding camps, or maker spaces. Support children in pursuing long-term projects aligned with their interests.

Balance Technology Use with Unplugged Time

While digital skills are important, balance is key. Set guidelines for screen time. Encourage outdoor activities and face-to-face social interactions. Promote reading physical books alongside digital content.

Develop Critical Thinking Skills

Prepare children to navigate complex information landscapes by asking thought-provoking questions and engaging in debates. Encourage analysis of media messages and advertising. Play strategy games that require logical thinking.

Remember, the goal is not to predict exactly what skills or knowledge your children will need, but to foster adaptability, curiosity, and resilience. By focusing on these areas, you are preparing your children not just for specific future scenarios, but for a lifetime of learning and growth.

Your role as a parent is crucial in shaping your children's future. By providing them with a transparent look into how you approach change and fostering an environment of continuous learning and exploration, you equip them with the tools to navigate an uncertain but exciting future.

The future of work and society may be unpredictable, but by cultivating these fundamental skills and mindsets, you are giving your children the best possible foundation to thrive in any future they may encounter. Your investment in their development today will pay dividends in their ability to adapt, innovate, and contribute meaningfully to society tomorrow.

6. Educational Institutions

The future of education demands a bold vision that embraces innovation while honoring core educational values, and your institution has the power to lead this transformation. By adopting the Fusion Model for innovation, you can systematically address challenges from the digital divide to AI integration, while creating an environment where experimentation and continuous learning flourish. The key lies in balancing technological advancement with human-centered learning—leveraging AI and adaptive platforms to personalize education while investing heavily in educator development and fostering essential skills like critical thinking, creativity, and emotional intelligence. Your leadership can reshape assessment methods, break down disciplinary silos, and forge meaningful industry partnerships that keep curricula relevant and dynamic.

As you implement these changes, remember that success depends on clear communication, comprehensive professional development, and a commitment to accessibility and inclusivity. The future classroom will likely blend physical and digital spaces, requiring robust systems for hybrid learning and a renewed focus on student agency and self-directed learning. While we can't predict exactly what tomorrow's workforce will demand, by creating an educational environment that cultivates curiosity, adaptability, and resilience, you're preparing both educators and students for lifelong learning and meaningful contribution to society. Your vision and commitment to thoughtful innovation today will shape not just your institution's future, but the very future of learning itself.

Education is Changing Because the World is Changing

The landscape of education is evolving at an unprecedented pace, driven by technological advancements, societal shifts, and changing workforce demands. To thrive in this dynamic environment, educational institutions must adopt a proactive stance, embracing innovation while maintaining a focus on core educational values. The Fusion Model for organizational adoption of innovation provides a valuable framework for navigating this transformation.

Begin with agenda setting. Clearly identify the challenges your institution faces and the opportunities for improvement. This might involve addressing the digital divide, preparing students for an AI-driven workforce, or reimagining assessment methods. Once you have set your agenda, move to the matching phase. Seek out innovations that align with your goals, whether they are technological solutions, pedagogical approaches, or organizational structures.

As you implement these innovations, enter the redefining and restructuring phase. Adapt new technologies and methodologies to fit your specific context. This might involve customizing AI-powered learning platforms to align with your curriculum or redesigning physical spaces to support more flexible, project-based learning. Remember, the goal is not to force-fit innovations into existing structures, but to thoughtfully reshape your institution to leverage these new tools and approaches effectively.

In the clarifying phase, focus on clear communication and professional development. Ensure all stakeholders understand the purpose and potential of new initiatives. Invest heavily in teacher training, recognizing that the success of any educational innovation ultimately depends on the educators implementing it. Create opportunities for educators to experiment, reflect, and share their experiences with colleagues.

Finally, work towards routinizing successful innovations. This involves integrating new practices into the fabric of your institution, making them a seamless part of your educational approach rather than isolated initiatives.

Throughout this process, keep several key principles in mind:

Embrace a culture of continuous learning and adaptation.

The pace of change in our world necessitates an educational approach that is flexible and responsive. Foster an environment where experimentation is encouraged and failure is viewed as a learning opportunity.

Focus on developing future-ready skills.

While content knowledge remains important, prioritize skills that will be valuable in any future scenario. These include critical thinking, problem-solving, creativity, digital literacy, emotional intelligence, and collaboration. Design learning experiences that explicitly develop these skills alongside traditional subject matter.

Leverage technology as a tool for personalization and engagement.

AI and adaptive learning platforms offer unprecedented opportunities to tailor education to individual student needs. However, remember that technology should enhance, not replace, human-centered learning. Strive for a balance where technology amplifies the impact of great teaching.

Reimagine assessment and credentialing.

Move beyond traditional standardized testing towards more holistic, competency-based assessment models. Explore the potential of blockchain for secure, verifiable micro-credentials that can more accurately reflect a learner's skills and knowledge.

Foster global connections and cultural competence.

Use technology to connect your students with peers around the world. Develop curricula that address global challenges, perhaps aligning with the United Nations Sustainable Development Goals.

Prioritize accessibility and inclusivity.

As education becomes increasingly technology-dependent, redouble efforts to bridge the digital divide. Ensure that innovations in educational technology do not exacerbate existing inequalities but instead serve to level the playing field.

Prepare for hybrid and flexible learning models.

The COVID-19 pandemic has accelerated the adoption of remote and hybrid learning. Develop robust systems that can seamlessly blend in-person and online learning experiences.

Cultivate partnerships with industry.

The rapid pace of technological change means that education must be more closely aligned with workforce needs. Develop strong relationships with industry partners to ensure your curricula remain relevant and to provide students with real-world learning opportunities.

Invest in educator development.

The role of educators is evolving from knowledge transmitters to learning facilitators. Provide ongoing professional development opportunities that empower educators to thrive in this new paradigm.

Emphasize interdisciplinary learning.

The complex challenges of the future will require the ability to integrate knowledge from multiple disciplines. Break down silos between departments and create opportunities for cross-disciplinary projects and courses.

Address ethical considerations proactively.

As AI and other emerging technologies become more prevalent in education, grapple with the ethical implications. Develop clear policies around data privacy, algorithmic bias, and the appropriate use of AI in assessment and instruction.

Foster student agency and self-directed learning.

Prepare students for a future where they will need to continuously upskill and reskill. Create opportunities for students to take ownership of their learning journey and develop the skills for lifelong learning.

Conclusion

In conclusion, the future of education is not about predicting exactly what knowledge or skills will be needed, but about creating learning environments that foster curiosity, adaptability, and resilience. By embracing innovation while staying true to core educational values, institutions can prepare students not just for the jobs of tomorrow, but for a lifetime of learning and meaningful contribution to society.

The journey of innovation in education is ongoing, and it requires leadership, vision, and a commitment to continuous improvement. By focusing on these key areas and leveraging the Fusion Model for innovation adoption, educational institutions can position themselves at the forefront of educational transformation, shaping the future of learning itself.

Summary

As we stand at the dawn of the Fourth Industrial Revolution, where digital, biological, and physical innovations converge to reshape our world, the future of education has never held more promise or importance. Through the lens of the Fusion Model, we can see how each stakeholder's role interweaves to create a tapestry of transformation that will define learning for generations to come.

For policymakers crafting flexible frameworks and administrators measuring return on instruction, your decisions ripple through the educational ecosystem, creating spaces where innovation can flourish while ensuring no learner is left behind. As you validate investments not just through traditional metrics but through the lens of improved student outcomes and teaching effectiveness, you're building the foundation for an education system that truly serves all.

To our educators at the forefront of change, your embrace of lifelong learning and dedication to developing student agency lights the way forward. By modeling adaptability and showing students how to navigate uncertainty with confidence—you are not just teaching subjects; you are empowering future innovators who will shape the world through their creativity, critical thinking, and emotional intelligence.

Students, you stand at a unique moment in history, where your journey of learning extends far beyond classroom walls. As you take ownership of your education, building digital literacy alongside human skills, and engaging in projects that cross traditional boundaries, you are not just preparing for the future—you are actively creating it. Your curiosity, adaptability, and entrepreneurial spirit will drive the innovations that define the Fourth Industrial Revolution.

Parents, your role in nurturing independence, balancing technological fluency with unplugged wisdom, and fostering a growth mindset is invaluable. By providing your children with both roots and wings—the stability of core values and the freedom to explore and adapt—you are helping shape a generation ready to face any challenge with resilience and creativity.

To educational institutions embracing continuous learning and adaptation, your commitment to reimagining assessment, fostering global connections, and building bridges with industry partners creates an ecosystem where innovation thrives. As you navigate the ethical implications of emerging technologies and prioritize accessibility, you're ensuring that education remains a powerful force for positive change in society.

Together, guided by the Fusion Model's framework of thoughtful innovation adoption, we are not just preparing for an uncertain future—we are actively shaping it into one of unprecedented opportunity. While we cannot predict exactly what knowledge tomorrow's world will demand, we can cultivate the curiosity, adaptability, and resilience that will allow us to face any challenge with confidence.

As digital, biological, and physical innovations continue to converge, our collective commitment to educational transformation becomes ever more crucial. Every stakeholder—from policymaker to parent, educator to student—plays a vital role in this journey. By embracing innovation while honoring our core educational values, we're creating learning environments that don't just react to change but drive positive transformation.

The future of education is not a distant horizon—it is being written today through our collective actions and choices. As we continue this journey together, let us remember that every step forward, every innovation embraced, and every learner empowered contributes to a future where education truly serves as the cornerstone of human progress. In this shared mission of transformation, we're not just preparing for the Fourth Industrial Revolution—we are ensuring that humanity's capacity to learn, adapt, and thrive evolves alongside our technological achievements, creating a future where everyone has the opportunity to reach their full potential and contribute meaningfully to our shared tomorrow.

Chapter 7: Preparing for the Future - Conclusion

Preparing for the Future

Recommendations by Stakeholder Group

Policymakers

Educational Paradigm Evolution

Technology Integration Framework

Funding & Resource Allocation

Data Privacy Guidelines

Educators

Innovation Adoption Mindset

Professional Development

Student Agency Support

AI Integration Strategies

Students

Develop Agency & Independence

Embrace Lifelong Learning

Build Digital Literacy

Practice Ethical Decision-Making

Parents

Support Agency Development

Balance Technology Use

Encourage Global Awareness

Foster Critical Thinking

Institutional Priorities

| Continuous Innovation | Equitable Access | Ethical Framework | Global Perspective |

HORIZON SCAN

As we prepare for the educational landscape of 2059, success depends on the coordinated efforts of multiple stakeholders, each playing their unique role in shaping the future of learning. From policymakers crafting flexible frameworks to educators fostering student agency, every participant in the educational ecosystem must embrace both innovation and human connection. The recommendations outlined in this chapter provide a roadmap for thoughtful transformation, emphasizing the importance of balancing technological advancement with human-centered learning principles. By adopting a strategic approach to change while maintaining focus on equity, accessibility, and ethical considerations, we can create an education system that truly serves all learners.

WAYPOINTS

Key Insight 1: Effective educational transformation requires coordinated action across multiple stakeholder groups - policymakers, administrators, educators, students, parents, and institutions

Key Insight 2: Return on Instruction (ROI) should be prioritized over traditional Return on Investment metrics when evaluating educational initiatives and technologies

Key Insight 3: Professional development and support for educators must evolve to match the changing demands of an AI-augmented learning environment

Key Insight 4: Student agency and self-directed learning capabilities become increasingly critical as we prepare learners for an uncertain future

NAVIGATION PROMPTS

◎ Implementation

- How might different stakeholders in your educational community collaborate more effectively to drive positive change?
- What strategies could help balance technological innovation with maintaining human connection in learning?

💡 Innovation

- How can your institution measure and optimize Return on Instruction when adopting new educational technologies?
- What professional development approaches might best prepare educators for an AI-augmented future?

🌐 Impact

- How might the recommended strategies affect different populations within your educational community?
- What steps could ensure that innovation efforts promote rather than hinder educational equity?

FUTURE LOG

Observations:

Ideas to Explore:

Next Steps:

CONCLUSION

SHAPING THE FUTURE OF LEARNING

1. Recap of key trends and potential futures

COVID19's impact on education showed us that we need to do a better job of preparing for our future. We were simply not ready from an infrastructure standpoint, and we learned from that. However, perhaps greater than that lesson, we learned that we desperately need each other. The learning journey must be humanized. The psychological impact and social impact on children and adults cannot be understated.

By 2029, we can expect to see artificial intelligence and machine learning playing a central role in personalizing education. AI-driven tutors will provide instant feedback and curriculum adjustments, tailoring the learning experience to each student's unique needs and pace. However, this integration of AI will not be without challenges. Ethical debates around AI's role in education will intensify, emphasizing the need for human oversight and the importance of "humanizing" AI in the learning process.

Virtual and augmented reality technologies will continue to break down geographical barriers, enabling global virtual classrooms and enhancing experiential learning through immersive simulations and field trips. These technologies will augment traditional learning materials, providing students with rich, interactive experiences that were previously impossible.

Looking further ahead to 2039, we may see the emergence of brain-computer interfaces in education. Non-invasive EEG-based interfaces could enhance focus and learning, with the potential for direct knowledge acquisition through neural implants. This development will undoubtedly raise ethical concerns around cognitive enhancement and educational equality.

The physical learning environment itself will evolve into a hyper-connected space, seamlessly integrating physical and virtual elements. These flexible, adaptable environments will support various teaching methods, though educators will need to carefully balance screen time with physical interaction and hands-on learning.

A shift from age-based to competency-based progression will redefine how we structure education. Personalized learning pathways will be tailored to individual needs and pace, focusing on skills acquisition rather than traditional subject-based curricula. This approach aligns with the growing emphasis on lifelong learning and the demands of the gig economy, where micro-credentials and modular learning options will become increasingly important.

The curriculum itself will become more globalized, with an increased focus on global awareness and cultural competency. Education will be designed around solving real-world problems, such as those outlined in the UN Sustainable Development Goals. This approach will integrate STEM thinking with humanities for holistic problem-solving, preparing students for the complex challenges of the future.

Blockchain technology will revolutionize credentialing, with blockchain-verified micro-credentials potentially replacing traditional degrees. This shift will place greater emphasis on skill verification and portfolios over institutional reputation, allowing for more flexible education models and easier combination of credentials from various sources.

The role of educators will undergo a significant transformation. Teachers will shift from being knowledge transmitters to learning facilitators, increasingly collaborating with AI teaching assistants. This evolution will necessitate continuous professional development in emerging technologies for educators to remain effective in their roles.

Accessible and inclusive education will be a key focus, with adaptive technologies addressing the needs of learners with disabilities and technological solutions helping to bridge socioeconomic barriers. Preserving linguistic and cultural diversity in a globalized educational landscape will be an ongoing challenge and priority.

Ethical considerations will be at the forefront of these changes. Issues such as data privacy, the commodification of learner information, and balancing efficiency with humanistic values in education will require careful navigation. The role of critical thinking in an AI-driven world will become more crucial than ever.

Environmental and sustainability education will be integrated more deeply into curricula, with technology being leveraged to teach environmental awareness and conservation. We may even see the emergence of space-based learning, with satellite-based internet improving global access to education and space simulations enhancing astronomy and earth science education.

To successfully adopt these innovations, educational institutions will need to progress through the stages outlined in the Fusion Model: agenda-setting, matching, redefining/restructuring, clarifying, and routinizing. This process will require a proactive approach to change management and a willingness to reimagine traditional educational structures and approaches.

As we navigate these trends, it is crucial to remember that the fundamental goal of education remains constant: to prepare learners for the challenges and opportunities of their future. By thoughtfully adopting innovations while preserving the core human elements of learning, we can create an education system that is both technologically advanced and deeply humanizing. The future of education is not just about implementing new technologies, but about fostering a culture of innovation and adaptability that will serve learners throughout their lives.

2. Access to Agency

As the world continues to change, education will continue to evolve. Yet it is important for educators to take a moment to reflect on teaching methods, rekindle the passion for education, and embrace change. We must learn to cultivate the courage to stay relevant by acknowledging that growth is an ongoing journey. By nurturing essential attributes of success: motivation, environment, skills, and knowledge we will empower ourselves and our students.

The courage to make change is at the heart of success. We can call this courage agency, that is, the feeling of being in control of your own learning. Agency is promoted in ourselves and others by vicarious experience, it is, therefore, more important than ever to transparently emphasize perseverance and hard work with our colleagues and students. We must explore, adopt, and inspire change in order to support a relevant educational experience, ensuring that our students are prepared for the challenges and opportunities of the future.

The future of work is being driven by innovation, and in education we must evolve to meet the needs of our society. We must seize the opportunity to adopt change, emerging technologies, and innovative teaching methods as we prepare our students for their future.

What do problem solvers, go-getters, and DIYers have in common? AGENCY

When we figure things out and/or get things done, we tend not to feel as though they simply happen to us, instead we feel as though we are in charge. The sense of agency refers to this feeling of being in the driving seat when it comes to our actions.[238] In an age where we see the best moments of people's lives through social media, we must be more transparent with our learners about the realness of life and the stick-to-it-iveness necessary to get things done. We need to provide access to our agency.

The Power of Access

We all know that access to a home library can have a powerful effect on children. My mother would frequently take us to the library to pick out whatever books we wanted. A parenting strategy she learned from her mother. These books provided me with access to dynamic places and opened doors for imagination, creativity, and critical thinking.

A study by Sikora, Evans, and Kelley (2019) on how access to books impact adult literacy, numeracy, and technology skills found that simply growing up in a home with enough books increases adult literacy and math prowess.

- A child growing up in a home with at least 80 books will

have greater literacy and numeracy in adulthood.

- A home library can promote reading and math skills more than college alone can.

- Growing up in a pro-learning home leads to a lifetime of knowledge-seeking. [239, 240]

But *access* does not stop there in education, access to emergent technology is very important. Our *approach* to emergent technologies in learning is very important. We do not talk about integrating reading into our teaching, we have adopted reading as a meaningful part of our teaching practice. The word "integration" connotes coercion without choice, whereas, the word "adoption" conveys agency and choice. We need to be cognizant of adopting innovative strategies and technology rather than simply integrating it. This conscious choice directly relates to the type and quality of access we provide our students to these tools.

Access to emergent technologies amplifies our skills

We know that closing the digital divide means providing access to emergent technologies, and the skills to leverage them, for our students and their future.

My dad always had computers around the house. I was never told I needed to learn coding or embrace this as the work tool of the future. I just had access. And that access has led to increased comfort around using computers on a daily basis.

We understand the benefits of having access to the internet as well, it increases educational and career opportunities. However, we still need to work to provide access to more people in what we call the "digital divide." While the digital divide continues, it has, by all indications, decreased in recent years. Measuring the number of people without access to computers and the internet does not fully describe the digital divide. Many researchers are now shifting from focusing merely on access to focusing on digital literacy.[241]

But access is not enough to bring about the paradigm shifts necessary to change education for the better.

Early research compared traditional classroom-based instruction to technology-supported instruction and found no significant benefit from adding technology to instruction.[242] This might seem counterintuitive to our arguments for the educational benefits of new technology, but a deeper dive into the research reveals important, productive insights: A particular technology used in classrooms is not as important as other instructional factors, such as pedagogy and course design. This is telling us what we already know, the importance of the teacher and that good teaching comes first. Our responsibility is to better understand emergent technologies as tools to improve learning and amplify good educational practices.

It is in the adoption, rather than integration, of emergent technologies that we will start to see a meaningful shift in practice. Simply integrating new technology will not change our practices. It's adopting a positive approach to innovation, and providing our learners with meaningful access to it, that will lead to a paradigm shift to better meet the needs of students in the future.

Agency

The benefits around access to books and closing the digital divide are clear. But there is more for us to do. We need to guide our students to a sense of agency by providing them with access to our own agency in action.

Agency is the capacity of individuals to act independently and make their own free choices, carrying out actions based on their desires and plans. This sense of agency, which is integral to cognitive development, empowers individuals to recognize themselves as independent and capable of controlling their own behavior.[243] People with a strong sense of agency are more likely to take on difficult tasks, set challenging goals, and persist in the face of setbacks. They also tend to think strategically, attribute failure to insufficient effort, and recover quickly from setbacks. In contrast, people who doubt their capabilities tend to shy away from difficult tasks, give up easily, and have low aspirations. They also tend to dwell on their deficiencies, focus on the negative consequences of failure, and undermine their own efforts by diverting attention from effective thinking and being slow to recover from setbacks.[244]

Positive Perceived Self-Efficacy toward our abilities is Agency.

Growing up, I would help my dad repair the family cars, everything from changing the breaks to swapping out engines. My role was to hand my dad the correct tools. We would sweat and toil together trudging through all kinds of grease and oil—yes, there are different kinds. I would not say that this taught me *how* to repair a car, but I learned that it could be done. I learned, through vicarious access to my Dad's agency, that I can use the skills I had to get things done—I could read a repair manual, I could use a tool… I learned that I CAN DO IT.

An increased sense of agency will foster an ethic of self-sufficiency, promoting the idea that anyone is capable of performing a variety of tasks for themselves.

Access to our agency will amplify our students preparedness for their future of work

We do not know all of the jobs that will exist for our students even in the relatively near future. We should not train our students to replicate our tasks that may soon be irrelevant. In fact, many of our more mundane tasks will be automated soon. The automation of tasks takes a lot of control out of our hands, simple things like auto-correct for typing and complex things like self-driving cars are examples of control headed toward automation. Automation raises concerns over agency, if we adopt technologies that automate, we are choosing to give up some control.[245] This is a bit scarier than you may think. For example, one area where automation is well-established is aircraft control, much of the pilot's work is carried out by a computer. A 2012 research study into the different levels of automation and the effects on pilots found that as automation increased, the sense of agency decreased (thankfully the research was done in flight simulators).[246] Studies like this remind us to be cognizant of our agency as it relates to future automation and how to be better prepared for the future.

Teach Agency in our own efforts to adopt emergent technology.

As professional educators, we want to deliver polished lessons where "nothing can go wrong," in effect, we are trying to deliver a social-media, picture-perfect lesson. The design of these lessons often leads us to a state of analysis paralysis, which actually prohibits us from trying new things. We must be transparent with our learners about the real-ness of life and the stick-to-it-iveness necessary to overcome obstacles and challenges in our everyday lives (even our teaching).

In an age where trying to master the latest and greatest is nearly impossible, our new norm is to innovate regularly. And this process must take place transparently in order to teach our students to prepare for a future that we can't even fully conceptualize. Trying new technology, investigating the relevance of a new tool, and having reflective dialogue around the value of what we are doing, must all be transparent. So that through vicarious experience and hands-on effort they will learn agency and amplify their preparedness for their future.

The importance of agency in successful edtech adoptions

A positive educational experience leveraging agency helps to ensure that students are ready for the challenges and opportunities of the future.

Education is changing because the world is changing. During the pandemic, teachers and students rapidly adopted new tools to pivot to remote and hybrid learning. With the advent of generative AI technologies, conversations in education are abuzz about AI's potential to redefine lesson delivery, homework, and formative assessments. The emergence of AI comes at an interesting time, when educators are looking for solutions to close the gap of pandemic-era learning loss and prepare their students for an increasingly technology-driven world.

As educators introduce new technologies in their classrooms, successful adoption and improvements to student outcomes will hinge on thoughtful strategies and intent to embrace change.

Advancements in education technology

History has taught us that innovation is inevitable. For instance, in classroom instruction, we have evolved from yesterday's chalkboards and overhead and LCD projectors to cutting-edge interactive displays. While educators can't predict the next breakthrough technology, we can anticipate the adoption process–by relating our past experiences to a positive path forward.

Instructional design and product development teams should consider the adoption process when designing products to meet the current needs of educators. Designers often pull from personal experiences and observations of how educators delivered lessons from the front of the classroom and created meaningful dialogue with students. For example, displays need to be simple and relatable for educators. The technology should augment what the teachers are already doing in the classroom and amplify their impact on student learning outcomes.

Educators and school leaders adopting new technologies—whether AI or hardware like interactive displays—need to understand their purpose. Do not purchase new technology because it is the hottest trend. Think about how you will leverage it to enrich the learning experience. For example, an LCD projector may have helped present lesson materials, but an interactive whiteboard can provide students with a new hands-on experience. They can understand the information more deeply, making connections they never would have before by looking at a static image. Innovation in technology and strategies can provide new opportunities to meet the diverse needs of our students.

Successful adoption also requires defining the measure of success. Higher test scores are often used to measure student achievement, but we know true meaning-making and application of skills are measured in different ways. A successful classroom has engaged students who are inspired to take ownership of their learning and become contributing members of their learning community. Educators attuned to their classroom culture understand when their students are engaged, and they seek to activate more and more of those coveted light-bulb moments—indicators of high motivation to learn more. The adoption of new technology will amplify instructional practice when used strategically to make a positive impact on students.

Access to agency

How should educators approach adopting a technology like AI where there is no past experience to serve as precedence? Teachers look for instructional leadership from their district, state or the Department of Education (DOE). In fact, the DOE's "Artificial Intelligence and the Future of Teaching and Learning" report recommends keeping humans in the loop with AI: "We envision a technology-enhanced future more like an electric bike and less like robot vacuums. On an electric bike, the human is fully aware and fully in control, but their burden is less, and their effort is multiplied by a complementary technological enhancement."[247] This high-level thinking is helpful, but more direction is needed.

However, guidance remains limited at the state level. The State Education Technology Directors Association (STEDA) reported in its 2023 State EdTech Trends that there is an "increased demand for guidance relating to artificial intelligence" and its use in K–12 schools, yet most states reported no efforts to meet that increased demand. Survey results showed that 55 percent of respondents indicated an increased interest in AI policy development—yet only two percent reported that their state has an AI initiative in place.[248]

How then are educators to adopt new technology like AI when AI's usage in education is not a question of "if" but "when"?

Technology is only going to permeate more of our everyday lives. We must equip students with valuable skills, such as how to write effective generative AI prompts, to prepare them for their future careers. Seek forward-thinking educators and industry experts online, where professional development, webinars, blogs and videos on AI are abundant. Importantly, do not be afraid of safely and thoughtfully experimenting with AI directly in front of your students. This is a difficult direction but remember: Young minds can be shaped by seeing trial and error. Rather than appear like we are the experts and already have all the answers, we can give our students visibility into the adoption process, including the struggles. This is called access to Agency.

Agency - the power to act, plan, and persevere - forms the foundation of human autonomy and growth. It manifests in our daily lives when we tackle challenges head-on, like teaching ourselves home repairs through online videos instead of immediately calling a professional. This self-directed problem-solving represents agency in its purest form, and it deserves a central place in modern education. By fostering students' sense of agency in the classroom, we cultivate essential capabilities like creative problem-solving, independent project execution, and collaborative thinking. These competencies are not just academic exercises - they are vital tools for navigating an evolving world where artificial intelligence and technological innovation continuously reshape the landscape of work and learning.

As the world continues to change, education will continue to evolve. Yet we need to take a moment to reflect on our educational systems, our instructional strategies, the tools we use and how each impacts effective teaching methods. In order to ignite a passion for lifelong learning that embraces change, we must cultivate the courage to stay relevant by acknowledging that technology is an evolving journey. By nurturing essential attributes of success, including motivation, environmental consideration, skills, and knowledge, we will empower ourselves and our students to adapt to breakthroughs.

The courage to make change is at the heart of success. We must explore, adapt, and inspire change to support a positive educational experience, ensuring that our students are ready for the challenges and opportunities of the future.[249]

3. You Cannot Be Replaced

My letter to teachers: You CANNOT be replaced… a lesson I learned from trophic cascades.

In one of my first years teaching, I was told that we could all be replaced. This was not a threat, nor was it directed at me personally, but offered as an insight into a perspective on the teaching profession. Young and full of energy, I fought this mentality, believing it untrue and that I had to simply work harder to make a tangible difference. As I got older I began to resign myself to this fact: I was a replaceable cog in a machine. I could still be a good cog, maybe even a great cog, but alas, a cog. What I failed to realize was the complex level of interconnectedness each of us has. I recently had a lightbulb moment when learning about connectivity as it relates to a scientific phenomenon called a trophic cascade.

Trophic Cascades

The addition or removal of top predators triggers a trophic cascade, an ecological process that results in significant changes

in ecosystem structure and nutrient cycling. American zoologist Robert Paine coined the term "trophic cascade" in 1980 to describe the cause-and-effect changes in food webs resulting from the experimental manipulation of top predators. The increase or decrease of carnivores in a food chain causes a reciprocal decrease or increase in herbivores and a corresponding increase or decrease in primary producers, such as plants and phytoplankton.[250]

The removal of wolves in North America led to an increase in white-tailed deer, which caused a decline in plants. This trophic cascade was exemplified in Yellowstone National Park, where the extermination of wolves between the 1880s and 1920s resulted in overgrazing by deer and elk, leading to a decrease in trees. This, in turn, loosened the soil on riverbanks, affecting beavers' ability to build dams and ultimately impacting fish populations. However, the reintroduction of gray wolves in the past decade has altered elk grazing patterns, allowing for the regrowth of trees and the gradual restoration of the ecosystem in Yellowstone.[251,252]

Some would like to ruin my overly-romanticized view of the ecosystem's symbiosis and how the reintroduction of wolves positively impacted Yellowstone, but we cannot deny that there is a real level of interconnectivity going on. While I do not propose each of us is a predator in the educational system, I will say we are having a real impact on those around us. Sometimes the impact is intangible and unseen, perhaps even seeming unmeasurable, but eventually these connections manifest themselves in many ways: the social media poke we get from a student we had in class decades ago or a direct message asking for help and/or reminding us of our shared experience (September 11 reminders always strikes a powerful chord with me). All of which stem from a level of connection we cannot see in the moment, a level of meaningful connection which says *you cannot be replaced.*

Perhaps you have gotten into a slump, resigning yourself to a cog-like mentality (I have been there too). You may not feel relevant, appreciated, or that you are making a difference. Now is the time to re-introduce *yourself to your WHY.* You matter, your work matters!

Also, please know, you are *not alone!* Wolves live and hunt in packs, these social animals cooperate and are known to roam large distances. Our Professional Learning Networks (PLN's) too are the embodiment of social cooperation that extends to the entire world! Reach out, share, and collaborate.

Note: The kindness in your eyes, the patience in your voice, etc., are the attributes of humanity that we need forever and always in educating our future.

4. Making a Meaningful Future

The future of education lies in our hands. As we stand at the brink of unprecedented technological and societal change, we must seize this moment to reimagine and reshape our educational landscape. Let us unite in our commitment to forge a path that nurtures curiosity, cultivates adaptability, and empowers every learner to thrive in an ever-evolving world.

Our call to action is clear: We must embrace innovation thoughtfully and strategically, leveraging the Fusion Model to guide our adoption of new technologies and methodologies. But let us remember that technology is a tool, not a solution in itself. Our focus must remain on developing the fundamental skills that will serve our students in any future scenario—critical thinking, creativity, emotional intelligence, and collaboration. We must reimagine assessment, moving beyond standardized tests to holistic, competency-based models that truly reflect a learner's capabilities. Let us harness the power of blockchain and micro-credentials to create more flexible, personalized learning pathways.

In our increasingly interconnected world, we have the opportunity—and the responsibility—to foster global awareness and cultural competence. Let us use technology to bridge distances, connecting our students with peers across the globe and tackling real-world challenges together. As we do so, we must remain vigilant in our commitment to accessibility and inclusivity, ensuring that technological advancements serve to level the playing field rather than widen existing gaps.

Educators, your role is evolving, but it remains central to this transformation. Embrace your position as facilitators of learning, guides in the vast sea of information. Invest in your own growth, continuously expanding your skills and understanding. Model the very adaptability and love for learning that we wish to instill in our students. And as we navigate this changing landscape, let us not forget the irreplaceable value of human connection. Technology should enhance, not replace, the powerful bond between teacher and student.

To all stakeholders in education—policymakers, administrators, teachers, parents, and students—this is our moment. We have the power to shape an educational system that not only prepares students for the jobs of tomorrow but equips them with the agility to thrive in a future we can scarcely imagine. Let us commit to fostering environments where curiosity flourishes, where failure is seen as a stepping stone to growth, and where every learner has the opportunity to reach their full potential.

The path ahead may be challenging, but it is filled with incredible possibilities. Together, we can create an education system that ignites passion, nurtures talent, and empowers the next generation to tackle the complex challenges of our world. The future of education—and indeed, the future of our society—depends on our collective action. Let us rise to this challenge with courage, creativity, and unwavering commitment. The time for transformation is now. Will you answer the call?

Chapter 8: Conclusion - Shaping the Future of Learning

Shaping the Future of Learning

Access to Agency and Making Education Meaningful

☌ Access to Agency

Technology as Tool for Empowerment

Skill Development Through Experience

Self-Directed Learning

Problem-Solving Mindset

⊘ Agency Development

Positive Self-Efficacy

Independent Decision Making

Growth Through Challenge

Resilience Building

◎ Future Preparedness

Adaptability Skills

Lifelong Learning Mindset

Technology Fluency

Global Citizenship

♀ Meaningful Learning

Real-World Application

Personal Connection

Community Impact

Purposeful Innovation

🔒 Keys to Transformation

| Empower Through Access | Foster Independence | Build Future Skills | Create Purpose |

HORIZON SCAN

The future of education lies not just in the technologies we adopt, but in how we empower humans to use them. Through our exploration of access to agency, we've seen that the most powerful educational transformations occur when we give learners visibility into our own processes of adaptation and growth. The trophic cascade metaphor reminds us that educators, like keystone species in an ecosystem, have far-reaching impacts that may not be immediately visible but are essential to the health of the entire educational environment. As we move toward 2059, our success will depend not on replacing human elements with technology, but on fostering the agency, resilience, and interconnectedness that make us uniquely human.

WAYPOINTS

Key Insight 1: Access to agency—letting students see how we navigate change and solve problems—is crucial for preparing them for an uncertain future

Key Insight 2: Like ecological systems, educational environments are complex networks where small changes can have far-reaching effects

Key Insight 3: Technology should amplify rather than replace human capabilities, serving as tools that enhance our agency rather than diminish it

Key Insight 4: The future requires educators who model lifelong learning, resilience, and transparent problem-solving for their students

NAVIGATION PROMPTS

🎯 Implementation

- How might you make your own learning process more visible to students?
- What opportunities exist in your educational setting to provide students with access to agency?

💡 Innovation

- How can you ensure technological innovations enhance rather than replace human connection in your educational context?
- What strategies could help maintain the balance between efficiency and humanization in learning?

🌐 Impact

- How might increased focus on agency affect different learners in your educational community?
- What ripple effects might occur from changing how we model learning and problem-solving?

FUTURE LOG

Observations:

Ideas to Explore:

Next Steps:

Appendix

Identification and Criteria for Five Stages of Organizational Adoption of Innovation

Stage Identification	Criteria
1. Agenda Setting: the organization identifies a problem.	1. Identify and prioritize needs and problems. 2. Search the organization's environment to locate innovations of potential usefulness to meet an organizational problem.
2. Matching: a problem from the organization's agenda is fit with an innovation, and this match is planned and designed.	1. Decision-makers determine the feasibility of the innovation in solving the organization's problem. 2. Decision-makers decide to accept or reject an innovation.
Decision to Adopt	
3. Redefining/Restructuring: an innovation has been adopted and now is re-invented to more closely fit within the organization's needs and structure.	1. Innovation is re-defined to explain how it can fit organizational needs. 2. The organization re-structures to fit the innovation to find it a home.

4. Clarifying: as the innovation is put into more widespread use, the idea gradually becomes clearer to the organization's members.	1. Innovation is employed within the organization. 2. More members of the organization seek to use the innovation.
5. Routinizing: the innovation becomes incorporated into the regular activities of the organization and loses its separate identity.	1. Innovation is a seamless part of daily operations. 2. The innovation's adoption is organization-wide.

Glossary

Agency: The capacity to act independently and make free choices. In education, agency refers to a learner's ability to take control of their learning journey and make decisions about their educational path. It is crucial for future success and reflects confidence in one's ability to exert control over motivation, behavior, and social environment.

ARCS Model: An instructional design model focusing on four key components: Attention, Relevance, Confidence, and Satisfaction. Used to promote and sustain motivation in the learning process.

Artificial Intelligence (AI): Technology designed to simulate human intelligence and cognitive processes. In education, AI is used to personalize learning experiences, provide immediate feedback, and assist with administrative tasks.

Blended Learning: An educational approach that combines traditional face-to-face classroom methods with online digital materials and interaction. This hybrid approach allows for both independent and collaborative learning experiences.

Blockchain: A decentralized digital ledger technology that creates an unchangeable record of transactions. In education, blockchain is particularly valuable for verifying credentials and creating secure, tamper-proof educational records.

Boolean Search: A type of search allowing users to combine keywords with operators (AND, OR, NOT) to produce more focused and productive results. Essential for effective digital research and information literacy.

Brain-Computer Interface (BCI): Technology that enables direct communication between the brain and external devices, typically computers. In education, BCIs may be used to enhance learning experiences and provide accessibility options for students with disabilities.

Collaborative Innovation: The practice of working together to create new solutions, particularly between humans and AI systems. This includes understanding how to effectively partner with AI tools while maintaining human creativity and critical thinking.

Developmental Significance: The recognition that age and developmental stage play crucial roles in learning processes, particularly in social learning and play-based activities.

Dual-coding Theory: A cognitive theory proposing that the human mind processes information through two separate channels: verbal and nonverbal. This theory emphasizes the importance of using both visual and verbal representations to enhance learning and memory.

Flipped Classroom: An instructional strategy where traditional learning activities, like lectures, are moved outside the classroom, while activities traditionally considered homework are moved into the classroom. This allows for more interactive and guided practice during class time.

Fourth Industrial Revolution: The current era characterized by the convergence of digital, biological, and physical technologies, particularly through the Internet of Things (IoT) and AI systems. This revolution is reshaping how we live, work, and learn.

Fusion Model: A framework for understanding and implementing organizational change, particularly in educational settings. It describes the stages organizations go through when adopting new innovations: agenda-setting, matching, redefining/restructuring, clarifying, and routinizing.

Holodeck: A reference to an immersive, virtual reality environment where learners can experience and interact with simulated scenarios for educational purposes.

Human Performance Technology (HPT): A systematic approach to improving productivity and competence in organizations. In education, HPT focuses on enhancing performance at individual, process, organizational, and societal levels.

Hybrid Learning: Similar to blended learning, but specifically referring to learning models that combine synchronous and asynchronous instruction, often with some students physically present while others participate remotely.

Individualized Education Programs (IEPs): Legal documents that outline the learning needs, services, and goals for students with disabilities in special education programs.

Internet of Things (IoT): A network of interconnected devices and objects that collect and exchange data. In education, IoT enables smart classrooms, automated attendance systems, and personalized learning environments.

Intuitive Decision Making: The process of making choices based on human insight and experience, particularly important when working with AI systems. It involves retaining human judgment to audit and validate AI outputs.

Large Language Models (LLMs): Advanced AI systems capable of understanding and generating human-like text. In education, LLMs can assist with content creation, tutoring, and personalized learning experiences.

Method of Loci: An ancient memory technique that involves associating information with specific locations in an imagined space. This technique has been shown to significantly enhance memory retention.

Micro-credentials: Digital certifications that verify specific skills or competencies. These are becoming increasingly important in education as alternatives to traditional degrees and certificates.

Mobile Learning (mLearning): Educational content and activities delivered through mobile devices. This approach enables learning anywhere, anytime, making education more accessible and flexible.

Moore's Law: The observation that computing power tends to double approximately every two years while costs decrease. In education, this implies continually increasing technological capabilities and accessibility.

Prompt Engineering: The skill of crafting effective inputs for AI systems to generate desired outputs. This includes understanding how to phrase questions and requests to get the most useful responses from AI tools.

Rational Decision Making: A logical, step-by-step approach to decision making, particularly important when working with AI systems. It involves critically evaluating whether data-driven conclusions align with human experience and needs.

SAMR Model: A framework for integrating technology into teaching (Substitution, Augmentation, Modification, Redefinition). This model helps teachers evaluate how technology integration transforms their teaching practice.

Schema: Mental frameworks that help organize and interpret information. In learning, schemas help students connect new information with existing knowledge.

Self-efficacy: An individual's belief in their capacity to execute behaviors necessary to produce specific performance attainments. It reflects confidence in one's ability to exert control over motivation, behavior, and social environment.

Skynet: A fictional artificial intelligence system from the Terminator films that becomes self-aware and turns against humanity. In educational discourse, Skynet is often referenced when discussing ethical concerns about AI development and the importance of maintaining human control over AI systems.

Societal Systems: The complex networks of human interaction and organization that make up our communities. In the context of AI and education, these systems require careful consideration to ensure technology serves human needs and values.

Spatial Memory Activation: The process of improving information recall by associating knowledge with specific physical locations. This concept is particularly relevant in classroom design and learning space organization.

Special Education: Specialized instruction designed to meet the unique needs of students with disabilities. This includes various support services and accommodations outlined in Individualized Education Programs (IEPs).

The Third Teacher: A concept that recognizes the physical environment as a crucial element in the learning process, alongside educators and peers. This perspective emphasizes the importance of thoughtful design in educational spaces.

Trophic Cascades: An ecological concept used metaphorically in education to illustrate how changes in one part of an educational system can have far-reaching effects throughout the entire system, similar to how changes in predator populations affect entire ecosystems.

Universal Wealth: A concept relating to the equitable distribution of resources and opportunities, particularly in educational contexts.

Upskilling: The process of learning new skills or improving existing ones, particularly in response to changing technological and workplace demands. This is becoming increasingly critical as the pace of change accelerates in the workplace.

Notes

Introduction

1   More specifically, philosophers expand on these capacities in the following ways:
1. Moral responsibility and ethical behavior: Immanuel Kant emphasized free will as crucial for moral responsibility in his works on ethics.
2. Personal identity and self-conception: John Locke discussed the importance of consciousness and agency in personal identity in his "Essay Concerning Human Understanding."
3. Human dignity and rights: Jean-Paul Sartre's existentialist philosophy emphasized human freedom as central to dignity and authenticity.
4. Creativity and innovation: Henri Bergson argued for a kind of free will that allows for genuine creativity in his work "Creative Evolution."
5. Social and legal systems: John Stuart Mill's "On Liberty" emphasized individual liberty and responsibility as foundations for social and political systems.

Chapter 1

2   James C. Mackenzie. (1894). The Report of the Committee of Ten. The School Review, 2(3), 146–155. http://www.jstor.org/stable/1074830

3   James C. Mackenzie. (1894). The Report of the Committee of Ten. The School Review, 2(3), 146–155. http://www.jstor.org/stable/1074830

4   Committee of Ten's Recommendations 1892. (2025). Knox.edu. http://faculty.knox.edu/jvanderg/202_K/Commof10Recom.htm

5   Perez, C. (2004). Technological revolutions, paradigm shifts and socio-institutional change. In E. Reinert (Ed.), Globalization, Economic Development and Inequality: An Alternative Perspective (pp. 217-242). Edward Elgar Publishing.

6   Stadler, C. (2024, September 6). The Generative AI Hype Is Almost Over. What's Next? Forbes.

7   Howell, J., Shea, C., & Higgins, C. (2005). Champions of product innovations: Defining, developing, and validating a measure of champion behavior. Journal of Business Venturing, 20(5), 641-661.

8   Greenhalgh, T., Robert, G., Macfarlane, F., Bate, P., & Kyriakidou, O. (2004). Diffusion Of Innovations In Service Organizations: Systematic Review And Recommendations. The Milbank Quarterly, 82(4), 581-629.

9   Darley, J. M. & Latané, B. (1968). 'Bystander intervention in emergencies: Diffusion of responsibility'. Journal of Personality and Social Psychology. 8: 377–383.

10 Shippee, M. (2019) WanderlustEDU: An Educator's Guide to Innovation, Change, and Adventure. San Diego: Dave Burgess Consulting, Inc.

11 Daily, Alisia (2014). 'Social Innovation and Innovation Champions: An Analysis of Public and Private Processes.' Diss. Virginia Commonwealth U, 2014.

12 Rogers, E.M. (2003). Diffusion of innovations (5th ed.). New York: Free Press.

13 Rogers, E.M. (2003). Diffusion of innovations (5th ed.). New York: Free Press.

14 Clarke, R. (1999, Sept). Roger Clarke's innovation diffusion theory.

15 Shippee, M. (2019) WanderlustEDU: An Educator's Guide to Innovation, Change, and Adventure. San Diego: Dave Burgess Consulting, Inc.

16 Shippee, Micah (2016) 'mLearning in the organizational innovation process.' Dissertation. https://surface.syr.edu/etd/452

17 Rogers, E.M. (2003). Diffusion of innovations (5th ed.). New York: Free Press.

18 Engeström, Y. (1987). Learning by Expanding: An activity-theoretical approach to developmental research. Helsinki: Orienta-Konsultit.

19 Rogers, E.M. (2003). Diffusion of innovations (5th ed.). New York: Free Press.

20 Rogers, E.M. (2003). Diffusion of innovations (5th ed.). New York: Free Press.

21 Shippee, M. (2019) WanderlustEDU: An Educator's Guide to Innovation, Change, and Adventure. San Diego: Dave Burgess Consulting, Inc.
22 Rogers, E.M. (2003). Diffusion of innovations (5th ed.). New York: Free Press.
23 Shippee, M. (2019) WanderlustEDU: An Educator's Guide to Innovation, Change, and Adventure. San Diego: Dave Burgess Consulting, Inc.
24 Ely, d. (1999) Conditions that Facilitate the Implementation of Educational Technology Innovations on JSTOR.
25 Bandura, A. (1977). Self-efficacy: Toward a unifying theory of behavioral change.
26 Ghaith, G. and Shaaban, K. (1999). The relationship between perceptions of teaching concerns, teacher efficacy, and selected teacher characteristics. Teaching and Teacher Education, 15, 487.
27 Rogers, E. M. (2003). Diffusion of innovations (5th ed.). New York: Free Press.
28 Moore, Gary C. and Benbasat, Izak. (1991). Development of an instrument to measure the perceptions of adopting an information technology innovation. Information Systems Research, 2(3), 192.
29 Gerber, S., & Scott, L. (2006) Designing a learning curriculum and technology's role in it. Educational Technology Research and Development, 55(1), 461-478.
30 Cuban, L. (1986). Teachers and Machines: The classroom use of technology since 1920. New York: Teachers College Press
31 Cuban, L. (1986). Teachers and Machines: The classroom use of technology since 1920. New York: Teachers College Press
32 Vardhan, H. (2002). Radio Broadcast Technology. Resonance, (January), 53–63.
33 Traub, D. (2004). The shift to seamless augmentation and 'humane' applications via mobile/wireless devices: a view to a future for lifelong learning.
34 Berman, S. D. (2008). The Return of Educational Radio? International Review of Research in Open and Distance Learning, 9(2), 1–8.
35 Romiszowski, A. J. (1974). The Selection and Use of Instructional Media. London, UK: Kogan Page Limited.
36 Bosch, A. (1997). Interactive Radio Instruction: Twenty-three years of improving education quality. Washington, DC: World Bank Group, 1(1).
37 Friend, J. (1989). Interactive Radio Instruction: developing instructional methods. British Journal of Educational Technology, 20(2), 106–114.
38 Dagron, G. (2001). Making Waves. New York, NY: Rockefeller Foundation.
39 Dede, C. (1996). The Evolution of Distance Education: Emerging Technologies and Distributed Learning. The American Journal of Distance Education, 10(2).
40 Romiszowski, A. J. (1974). The Selection and Use of Instructional Media. London, UK: Kogan Page Limited.
41 Berman, S. D. (2008). The Return of Educational Radio? International Review of Research in Open and Distance Learning, 9(2), 1–8.
42 Cuban, L. (1986). Teachers and Machines: The classroom use of technology since 1920. New York: Teachers College Press.
43 Shiels, M. (2003). A chat with the man behind mobiles. BBC News.
44 Dede, C. (1996). The Evolution of Distance Education: Emerging Technologies and Distributed Learning. The American Journal of Distance Education, 10(2).
45 Cuban, Larry. Teachers and Machines: the Classroom Use of Technology since 1920. Teachers College Press, 2004.

Chapter 2

46 Ormrod, J. (2019). Human learning (8th ed.). Upper Saddle River, NJ: Pearson.
47 Collins, A. (2006). Cognitive apprenticeship. In R.K. Sawyer (ed.), The Cambridge handbook of the learning sciences (pp. 47-60). Cambridge, England: Cambridge University Press.
48 Collins, A., Brown, J.S., & Newman, S.E. (1989). Cognitive apprenticeship: Teaching the crafts of reading, writing, and mathematics. In L.B. Resnick (Ed.), Knowing, learning, and instruction: Essays in honor of Robert Glaser (pp. 453-494). Hillsdale, NJ: Erlbaum

Chapter 3

49 Harless, J.H. (1973). An analysis of front-end analysis. Improving Human Performance: A Research Quarterly, 4, 229-244.

50 Leijon, M., Nordmo, I., Tieva, Å., & Troelsen, R. (2022). Formal learning spaces in Higher Education – a systematic review. Teaching in Higher Education, 29(6), 1460–1481.

51 Leijon, M., Nordmo, I., Tieva, Å., & Troelsen, R. (2022). Formal learning spaces in Higher Education – a systematic review. Teaching in Higher Education, 29(6), 1460–1481.

52 Leijon, M., Nordmo, I., Tieva, Å., & Troelsen, R. (2022). Formal learning spaces in Higher Education – a systematic review. Teaching in Higher Education, 29(6), 1460–1481.

53 Barrett, P.S., & Zhang, Y. (2009). Optimal learning spaces: Design and implications for primary schools Salford University.

54 Leijon, M., Nordmo, I., Tieva, Å., & Troelsen, R. (2022). Formal learning spaces in Higher Education – a systematic review. Teaching in Higher Education, 29(6), 1460–1481.

55 Niemi, K. (2020). 'The best guess for the future?' Teachers' adaptation to open and flexible learning environments in Finland. Education Inquiry, 12(3), 282–300.

56 Leijon, M., Nordmo, I., Tieva, Å., & Troelsen, R. (2022). Formal learning spaces in Higher Education – a systematic review. Teaching in Higher Education, 29(6), 1460–1481.

57 Leijon, M., Nordmo, I., Tieva, Å., & Troelsen, R. (2022). Formal learning spaces in Higher Education – a systematic review. Teaching in Higher Education, 29(6), 1460–1481.

58 Leijon, M., Nordmo, I., Tieva, Å., & Troelsen, R. (2022). Formal learning spaces in Higher Education – a systematic review. Teaching in Higher Education, 29(6), 1460–1481.

59 Sørensen, Estrid. 2009. The Materiality of Learning: Technology and Knowledge in Educational Practice. New York: Cambridge University Press

60 Mulcahy, Dianne. 2018. 'Assembling Spaces of Learning 'In' Museums and Schools: A Practice-based Sociomaterial Perspective.' In Spaces of Teaching and Learning. Understanding Teaching-learning Practice, edited by R. Ellis and P. Goodyear, 13–29. Singapore: Springer

61 Schemas - The Decision Lab https://thedecisionlab.com/reference-guide/psychology/schemas

62 [5] What is a Schema in Psychology? - Structural Learning https://www.structural-learning.com/post/schema-in-psychology

63 Schemas - The Decision Lab https://thedecisionlab.com/reference-guide/psychology/schemas

64 [6] Schemas and Decision Making - SpringerLink https://link.springer.com/10.1007%2F978-1-4419-1428-6_264

65 Schemas - The Decision Lab https://thedecisionlab.com/reference-guide/psychology/schemas

66 How Does Schema Influence Your Customer's Decisions? https://trainingindustry.com/blog/sales/how-does-schema-influence-your-customers-decisions/

67 Schemas - The Decision Lab https://thedecisionlab.com/reference-guide/psychology/schemas

68 Schemas: the good, the bad and the reconstructed https://apsychologyteacherwrites.wordpress.com/2022/05/14/schemas-the-good-the-bad-and-the-reconstructed/

69 Schemas - The Decision Lab https://thedecisionlab.com/reference-guide/psychology/schemas

70 Schemas: the good, the bad and the reconstructed https://apsychologyteacherwrites.wordpress.com/2022/05/14/schemas-the-good-the-bad-and-the-reconstructed/

71 Schemas: the good, the bad and the reconstructed https://apsychologyteacherwrites.wordpress.com/2022/05/14/schemas-the-good-the-bad-and-the-reconstructed/

72 Bruning, K. (2023, December 9). 10 Reasons Zune Just Couldn't Keep Up With The iPod. SlashGear.

73 Schemas and Decision Making - SpringerLink https://link.springer.com/10.1007%2F978-1-4419-1428-6_264

74 Schemas: the good, the bad and the reconstructed https://apsychologyteacherwrites.wordpress.com/2022/05/14/schemas-the-good-the-bad-and-the-reconstructed/

75 Auddie Mastroleo. (2024, July 25). Using Juicy Sentences to Explore Complex Texts in ELA and Beyond. Edutopia; George Lucas Educational Foundation.

76 Royle, O. R. (2024, November). Gen Z and young millennial employees are missing the equivalent of one day's work every week due to mental health concerns, research shows.

77 Bandura, A. (1977). Social learning theory. Prentice-Hall.

78 Meltzoff, A. N., & Moore, M. K. (1977). Imitation of facial and manual gestures by human neonates. Science, 198(4312), 75-78.

79 Saarni, C., Mumme, D. L., & Campos, J. J. (1998). Emotional development: Action, communication, and understanding. Handbook of child psychology, 5, 237-309.

80 Whitebread, D., & Bingham, S. (2014). Habit formation and learning in young children. Money Advice Service.

81 Meltzoff, A. N., & Moore, M. K. (1977). Imitation of facial and manual gestures by human neonates. Science, 198(4312), 75-78.

82 Frith, C. D., & Frith, U. (2012). Mechanisms of social cognition. Annual review of psychology, 63, 287-313.

83 Brechwald, W. A., & Prinstein, M. J. (2011). Beyond homophily: A decade of advances in understanding peer influence processes. Journal of Research on Adolescence, 21(1), 166-179.

84 Steinberg, L., & Monahan, K. C. (2007). Age differences in resistance to peer influence. Developmental psychology, 43(6), 1531-1543.

85 Erikson, E. H. (1968). Identity: Youth and crisis. W. W. Norton & Company.

86 Kroger, J. (2007). Identity development: Adolescence through adulthood. Sage Publications.

87 Prinstein, M. J., & Dodge, K. A. (2008). Understanding peer influence in children and adolescents. Guilford Press.

88 Bandura, A. (1986). Social foundations of thought and action: A social cognitive theory. Prentice-Hall.

89 Mezirow, J. (1997). Transformative learning: Theory to practice. New directions for adult and continuing education, 1997(74), 5-12.

90 Manuti, A., Pastore, S., Scardigno, A. F., Giancaspro, M. L., & Morciano, D. (2015). Formal and informal learning in the workplace: a research review. International Journal of Training and Development, 19(1), 1-17.

91 Kuijk, J. V., Prehn, B., & Grong, A. (2020). Social Learning in the Workplace: A Perspective From Adult Learning Theory. In Competence Perspectives on Managing Internal Processes (pp. 139-160). Palgrave Macmillan, Cham.

92 Pasupathi, M. (1999). Age differences in response to conformity pressure for emotional and non-emotional material. Psychology and Aging, 14(1), 170-174.

93 Blanchard-Fields, F., & Horhota, M. (2005). Age differences in the correspondence bias: When a crime is more than a crime. The Journals of Gerontology Series B: Psychological Sciences and Social Sciences, 60(5), 279-288.

94 Flavell, J. H., Miller, P. H., & Miller, S. A. (2002). Cognitive development. Prentice Hall.

95 Frith, C. D., & Frith, U. (2012). Mechanisms of social cognition. Annual review of psychology, 63, 287-313.

96 Saarni, C., Mumme, D. L., & Campos, J. J. (1998). Emotional development: Action, communication, and understanding. Handbook of child psychology, 5, 237-309.

97 Kifer, Y., Buckingham, J., Sahin, H., Camstra, A., Kashte, S., Mauthe-Kaddoura, A., ... & Sarna, D. (2013). Social cognitive skills in children with learning disabilities. Journal on Educational Psychology, 7(2), 23-35.

98 Bandura, A. (1977). Social learning theory. Prentice-Hall.

99 Gosso, Y., Otta, E., Ribeiro, F. D. J. L., Bussab, V. S. R., & Almeida, H. J. F. D. (2007). Play in hunter–gatherer society. In Origins of the modern mind (pp. 213-253). University of Chicago Press.

100 Lillard, A. S. (2013). Playful learning and Montessori education. American Journal of Play, 5(2), 157-186.

101 Piaget, J. (1962). Play, dreams and imitation in childhood. W. W. Norton & Company.

102 Vygotsky, L. S. (1978). Mind in society: The development of higher psychological processes. Harvard University Press.

Chapter 4

103 BenBassett, D.. (2016). Linkedin.com.

104 National Education Association (2010). Preparing 21st Century Students for a Global Society: An Educator's Guide to the 'Four Cs'.

105 Industrial Revolution | Definition, History, Dates, Summary, & Facts | Britannica Money. (2023). In Encyclopedia Britannica.

106 Schwab, K., & World Economic Forum. (2016, January 14). The Fourth Industrial Revolution: what it means and how to respond.

107 Shippee, M. (2021) A Globalized Curriculum: The Next Evolution Of Education. International Conferences Mobile Learning 2021 (ML 2021) and Educational Technologies 2021 (ICEduTech 2021)

108 Muhammad Anshari, & Hamdan, M. (2022). Understanding knowledge management and up-skilling in Fourth Industrial Revolution: transformational shift and SECI model.

109 Wilson, J. H. and Daugherty, P.R. Embracing Gen AI at Work. (2024, September). Harvard Business Review.

110 Unskilled Labor: Definition and Examples of Jobs To Pursue. (2024). Indeed Career Guide.

111 Artificial Intelligence and the Future of Teaching and Learning - Office of Educational Technology. (2023, May 24). Office of Educational Technology.

112 Zaman, R. (2020, July 15). Unfolding Fourth Industrial Revolution - THE WAVES.

113 Shippee, M. (2023, December 22). Beyond the 4C's: Framing our Understanding of Future Skills in the era of the Fourth Industrial Revolution.

114 Shippee, M. (2023, November 30). AI in Education: a focus on pedagogy.

115 Shippee, Micah, The Fusion Model for Organizational Adoption of Innovation (May 4, 2020). http://dx.doi.org/10.2139/ssrn.4664783

116 Abdulkader, S. N., et al. (2015). Brain computer interfacing: Applications and challenges. Egyptian Informatics Journal, 16(2), 213-230.

117 Jain, V. K., et al. (2021). 'Satellite Internet constellation: A review of technology and launch of Starlink.' International Journal of Science and Research, 10(4), 1108-1114.

118 Vergara, D., et al. (2017). 'Virtual reality in space science education.' International Journal of Virtual and Augmented Reality, 1(2), 25-37.

119 Lion-Bailey, C., Lubinsky, J., and Shippee, M. (2020) Reality Bytes: Innovative Learning Using Augmented and Virtual Reality. San Diego: Dave Burgess Consulting, Inc.

120 M. Shippee and J. Lubinsky, 'Training and Learning in Virtual Reality: Designing for Consistent, Replicable, and Scalable Solutions,' 2021 International Conference on Electrical, Computer and Energy Technologies (ICECET)

121 Lion-Bailey, C., Lubinsky, J., & Shippee, M. (2022) The XR ABC Framework: Fostering Immersive Learning Through Augmented and Virtual Realities.

122 Shuell, T. J., & Farber, S. L. (2001). 'Students as co-investigators of data from space science missions.' Journal of Science Education and Technology, 10(4), 351-359.

123 Prisk, G. K. (2019). 'Microgravity and the lung.' Journal of Applied Physiology, 126(5), 1456-1465.

124 Baker, V. R. (2001). 'Water and the martian landscape.' Nature, 412(6843), 228-236.

125 Impey, C. (2020). 'Astrobiology and society: building an interdisciplinary research community.' Journal of Astrobiology, 19(6), 751-756.

126 Petroni, G., et al. (2019). 'Space technology transfer: Spin-off cases from space to Earth.' International Journal of Engineering Business Management, 11, 1847979019874409.

127 Razzaq, S., et al. (2019). 'Space education and outreach: Current status and future prospects.' Space Policy, 47, 194-206.

128 Webber, D. (2013). 'Space tourism: Its history, future and importance.' Acta Astronautica, 92(2), 138-143.

129 Schwartz, J. S., & Milligan, T. (Eds.). (2016). The ethics of space exploration. Springer International Publishing.

130 Haviland, W.A., Prins, H.E.L., Walrath, D, and McBride, B.. Cultural Anthropology: The Human Challenge., 2005.

131 Shippee, M. (2019) WanderlustEDU: An Educator's Guide to Innovation, Change, and Adventure. San Diego: Dave Burgess Consulting, Inc.

132 Cuban, L. (1986). Teachers and Machines: The classroom use of technology since 1920. New York: Teachers College Press.

133 Cuban, L. (1986). Teachers and Machines: The classroom use of technology since 1920. New York: Teachers College Press.

134 Frick, K. T. (2017). The Cost of the Technological Sublime: Daring Ingenuity and the new San Francisco-Oakland Bay Bridge.

135 Popkewitz, T. (1984). Paradigm and ideology in educational research : the social functions of the intellectual. London New York: Falmer Press.

136 Van Doren, C. (1992). A history of knowledge: past, present, and future. New York, N.Y: Ballantine Books.

137 Lahav, R. (1973). Futurology and Education: Four Futurologists and Their Theories of Education.

138 Toffler, A. (1970). Future shock. New York: Random House.

139 Van Doren, C. (1992). A history of knowledge: past, present, and future. New York, N.Y: Ballantine Books.

140 Freeman, B., Marginson, S., and Tytler, R. (2019). An international view of STEM education.

141 National Academy of Sciences, National Academy of Engineering, & Institute of Medicine of the National Academies. (2007). Rising above the gathering storm: Energizing and employing America for a brighter economic future.

142 National Academy of Sciences, National Academy of Engineering, & Institute of Medicine of the National Academies. (2007). Rising above the gathering storm: Energizing and employing America for a brighter economic future.

143 Carpenter, E., Regassa, L.B. and Watson, C.L. 'Accelerating Discovery: Educating the Future STEM Workforce | NSF - National Science Foundation.'

144 Akerson, V. L., Burgess, A., Gerber, A., Guo, M., Khan, T. A., & Newman, S. (2018). Disentangling the Meaning of STEM: Implications for Science Education and Science Teacher Education. Journal of Science Teacher Education, 29(1), 1–8. https://doi.org/10.1080/1046560x.2018.1435063

145 sabbott. (2013, December 3). Hidden Curriculum Definition. The Glossary of Education Reform. https://www.edglossary.org/hidden-curriculum/#:~:text=The%20hidden%2Dcurriculum%20concept%20is,of%20people%3B%20or%20what%20ideas

146 Micro-Credentials | NEA. (2024). Nea.org. https://www.nea.org/professional-excellence/professional-learning/micro-credentials

147 Gent, E. (2020, February 11). Is education the new currency? Bbc.com; BBC. https://www.bbc.com/worklife/article/20200210-is-education-the-new-currency

148 Micro-Credentials | NEA. (2024). Nea.org. https://www.nea.org/professional-excellence/professional-learning/micro-credentials

149 THE 17 GOALS | Sustainable Development. (2015). Un.org. https://sdgs.un.org/goals

150 THE 17 GOALS | Sustainable Development. (2015). Un.org. https://sdgs.un.org/goals

151 Merrill, D.M. 'First principles of instruction.' Educational Technology Research and Development.

152 Shippee, M. (2016). mLearning in the organizational innovation process. Dissertation. Syracuse University. Syracuse, NY

153 Shippee, M. (2019) WanderlustEDU: An Educator's Guide to Innovation, Change, and Adventure. San Diego: Dave Burgess Consulting, Inc.

154 Lubinsky, J. and Shippee, M (2021) Esports and strategic organizational adoption of innovation. (unpublished)

155 THE 17 GOALS | Sustainable Development. (2015). Un.org. https://sdgs.un.org/goals

156 THE 17 GOALS | Sustainable Development. (2015). Un.org. https://sdgs.un.org/goals

157 Bybee, R. W. The Case for STEM Education: Challenges and Opportunities. Arlington, Virginia: National Science Teachers Association, 2013.

158 Roxby, P. (2015, May 17). Why an iron fish can make you stronger. BBC News. https://www.bbc.com/news/health-32749629

159 Gartner Hype Cycle Reveals Top Technologies That Will Transform Sales. (2024). Gartner. https://www.gartner.com/en/newsroom/press-releases/2024-08-27-gartner-hype-cycle-reveals-top-technologies-that-will-transform-sales-in-the-next-decade

160 Holder, S. (2021, December 22). Pandemic Remote Work Spawns a New Class of Office Super Commuters. Bloomberg.com; Bloomberg. https://www.bloomberg.com/news/articles/2021-12-22/pandemic-remote-work-spawns-a-new-class-of-office-super-commuters

161 (2019). DataReportal – Global Digital Insights. DataReportal – Global Digital Insights. https://datareportal.com/global-digital-overview

162 Nations, U. (2018). New Technologies and the Global Goals | United Nations. United Nations.

https://www.un.org/en/un-chronicle/new-technologies-and-global-goals

163   Gender Gap - Mobile for Development. (2019). Mobile for Development. https://www.gsma.com/r/gender-gap/

164   Shippee, Micah, "mLearning in the organizational innovation process" (2016). Dissertations - ALL. 452.
https://surface.syr.edu/etd/452

165   Gill, S. (2025). How Many People Own Smartphones in the World? (2024-2029) | Priori Data. Priori Data. https://prioridata.com/data/smartphone-stats/

166   Forecast number of mobile users worldwide 2020-2025 | Statista. (2020). Statista. https://www.statista.com/statistics/218984/number-of-global-mobile-users-since-2010/

167   Handley, L. (2019, January 24). Nearly three quarters of the world will use just their smartphones to access the internet by 2025. CNBC. https://www.cnbc.com/2019/01/24/smartphones-72-per-cent-of-people-will-use-only-mobile-for-internet-by-2025.html#:~:text=Nearly%20three%20quarters%20of%20the,access%20the%20internet%20by%202025&text=Almost%20three%20quarters%20(72.6%20percent,to%20nearly%203.7%20billion%20people

168   Innovating to Keep Kids Learning During the Pandemic in Sri Lanka. (2021). UNICEF USA. https://www.unicefusa.org/stories/innovating-keep-kids-learning-during-pandemic-sri-lanka

169   (2021, December 13). 5 Ways to Give to Children in Need This Season With UNICEF. https://www.unicefusa.org/stories/5-ways-give-children-need-season-unicef

170   Wonolo. (2018, February 15). Data on the Gig Economy and How it is Transforming the Work-force. Wonolo. https://info.wonolo.com/blog/data-gig-economy-transforming-workforce/

171   Tunkel, A. (2018, August 13). Three Trends On The Future Of Work. Forbes. https://www.forbes.com/sites/forbesbusinessdevelopmentcouncil/2018/08/13/three-trends-on-the-future-of-work/

172   Rouse M. (2019). Gig economy. WhatIs.com. Retrieved from https://whatis.techtarget.com/definition/gig-economy

173   Cambridge Dictionary. (2025, February 12). gig economy. @CambridgeWords. https://dictionary.cambridge.org/us/dictionary/english/gig-economy

174   Dual Enrollment: Participation and Characteristics. (2019, February 5). Ed.gov.

175   2020 OCW Impact Report | Open Learning. (2020). Mit.edu.

176   Dual Enrollment: Participation and Characteristics. (2019, February 5). Ed.gov.

177   Shippee, M. & Lubinsky (2021) Training and Learning in Virtual Reality: Designing for Consistent, Replicable, and Scalable Solutions.

Chapter 5

178   Zheng, Z., Xie, S., Dai, H., Chen, X., & Wang, H. (2017). An overview of blockchain technology: Architecture, consensus, and future trends.

179   Grech, A., & Camilleri, A. F. (2017). Blockchain in education. Luxembourg: Publications Office of the European Union.

180   LinkedIn Learning: Online Courses for Creative, Technology, Business Skills. (2024).

181   HarvardX. (2024). Harvard.edu.

182   Gräther, W., Kolvenbach, S., Ruland, R., Schütte, J., Torres, C., & Wendland, F. (2018). Blockchain for education: lifelong learning passport.

183   Rooksby, J., & Dimitrov, K. (2019). Trustless education? A blockchain system for university grades.

184   Ocheja, P., Flanagan, B., & Ogata, H. (2018). Connecting decentralized learning records: a blockchain based learning analytics platform.

185   Chen, G., Xu, B., Lu, M., & Chen, N. S. (2018). Exploring blockchain technology and its potential applications for education.

186   Micro-Credentials | NEA. (2024). Nea.org. https://www.nea.org/professional-excellence/professional-learning/micro-credentials

187   Harless, J.H. (1973). An analysis of front-end analysis. Improving Human Performance: A Re-

search Quarterly, 4, 229-244.

188   Discover your Hogwarts house on Pottermore | Wizarding World.'

189   Mcalpine, Fraser. 'Happy Birthday, T.S. Eliot: 20 of His Most Life-Affirming Quotes | BBC America.'

190   Thornburg, D. D. (2013). From the campfire to the holodeck: Creating engaging and powerful 21st century learning environments. John Wiley & Sons, Oct 21, 2013

191   Bowie, A. (2023, February 15). 3 Key Components of Early Childhood Learning Environments. MiEN   Company®.   https://miencompany.com/3-key-components-of-early-childhood-learning-environments/

192   Bowie, A. (2023, February 15). 3 Key Components of Early Childhood Learning Environments. MiEN   Company®.   https://miencompany.com/3-key-components-of-early-childhood-learning-environments/

193   Bowie, A. (2023, February 15). 3 Key Components of Early Childhood Learning Environments. MiEN   Company®.   https://miencompany.com/3-key-components-of-early-childhood-learning-environments/

194   Sergiy Movchan. (2024, February 25). Good Learning Environments: Key Features Explained. Raccoon Gang. https://raccoongang.com/blog/what-makes-good-learning-environment/

195   Ibrahim, M., & Osama Al-Shara. (2007). Impact of Interactive Learning on Knowledge Retention. Lecture Notes in Computer Science, 347–355. https://doi.org/10.1007/978-3-540-73354-6_38

196   Sergiy Movchan. (2024, February 25). Good Learning Environments: Key Features Explained. Raccoon Gang. https://raccoongang.com/blog/what-makes-good-learning-environment/

197   Sergiy Movchan. (2024, February 25). Good Learning Environments: Key Features Explained. Raccoon Gang. https://raccoongang.com/blog/what-makes-good-learning-environment/

198   Sergiy Movchan. (2024, February 25). Good Learning Environments: Key Features Explained. Raccoon Gang. https://raccoongang.com/blog/what-makes-good-learning-environment/

199   Huang, R., Yang, J., Zheng, L. (2013) The Components and Functions of Smart Learning Environments for Learning. International Journal for Educational Media and Technology

2013, Vol.7, No. 1, pp. 4-14. https://jaems.jp/contents/icomej/vol7/IJEMT7.4-14.pdf

200   Huang, R., Yang, J., Zheng, L. (2013) The Components and Functions of Smart Learning Environments for Learning. International Journal for Educational Media and Technology

2013, Vol.7, No. 1, pp. 4-14. https://jaems.jp/contents/icomej/vol7/IJEMT7.4-14.pdf

201   Rusticus, S. A., Pashootan, T., & Mah, A. (2022). What are the key elements of a positive learning environment? Perspectives from students and faculty. Learning Environments Research, 26(1), 161-175. https://doi.org/10.1007/s10984-022-09410-4

202   Lion-Bailey, C., Lubinsky, J., and Shippee, M. (2020) Reality Bytes: Innovative Learning Using Augmented and Virtual Reality. San Diego: Dave Burgess Consulting, Inc.

203   Lion-Bailey, C., Lubinsky, J., and Shippee, M. (2020) Reality Bytes: Innovative Learning Using Augmented and Virtual Reality. San Diego: Dave Burgess Consulting, Inc.

204   Cline, Ernest. Ready Player One. Broadway Books, 2011

205   Bandura, A. (1977). Self-Efficacy: Toward a Unifying Theory of Behavioral Change. Psychological Review 84 (2), 191-215.

206   Bandura, A. (1977). Self-Efficacy: Toward a Unifying Theory of Behavioral Change. Psychological Review 84 (2), 191-215.

207   Bandura, A. (1977). Self-Efficacy: Toward a Unifying Theory of Behavioral Change. Psychological Review 84 (2), 191-215.

208   Bandura, A. (1977). Self-Efficacy: Toward a Unifying Theory of Behavioral Change. Psychological Review 84 (2), 191-215.

209   Cline, Ernest. Ready Player One. Broadway Books, 2011

210   Q and A: Imagining a Virtual Education Oasis.' Education Week Digital Directions. 8 Feb. 2012.

211   Shippee, M. Next Steps for AR and VR in Education: Harnessing The Power of Shared-Experience #edtech. (2017, December 14).

212   Mayer, R. E., & Moreno, R. (2003). Nine ways to reduce cognitive load in multimedia learning. Educational Psychologist, 38(1), 43-52.

213 Paivio, A. (1991). Dual coding theory: Retrospect and current status. Canadian Journal of Psychology, 45(3), 255-287.

214 Traub, D. (2004). The shift to seamless augmentation and 'humane' applications via mobile/wireless devices: a view to a future for lifelong learning.

215 Vardhan, H. (2002). Radio Broadcast Technology. Resonance, (January), 53–63.

216 Berman, S. D. (2008). The Return of Educational Radio? International Review of Research in Open and Distance Learning, 9(2), 1–8.

217 Romiszowski, A. J. (1974). The Selection and Use of Instructional Media. London, UK: Kogan Page Limited.

218 Berman, S. D. (2008). The Return of Educational Radio? International Review of Research in Open and Distance Learning, 9(2), 1–8.

219 Bosch, A. (1997). Interactive Radio Instruction: Twenty-three years of improving education quality. Washington, DC: World Bank Group, 1(1).

220 Friend, J. (1989). Interactive Radio Instruction: developing instructional methods. British Journal of Educational Technology, 20(2), 106–114.

221 Romiszowski, A. J. (1974). The Selection and Use of Instructional Media. London, UK: Kogan Page Limited.

222 Toward the Stars: Technological, Ethical, and Sociopolitical Dimensions of Interstellar Exploration. (2024). Arxiv.org.

Chapter 6

223 Autor, D. H. (2015). Why are there still so many jobs? The history and future of workplace automation. Journal of Economic Perspectives, 29(3), 3-30.

224 Brynjolfsson, E., & McAfee, A. (2014). The second machine age: Work, progress, and prosperity in a time of brilliant technologies. W. W. Norton & Company.

225 Kolarevic, B. (2003). Architecture in the digital age: Design and manufacturing. Taylor & Francis.

226 Sparrow, B., Liu, J., & Wegner, D. M. (2011). Google effects on memory: Cognitive consequences of having information at our fingertips. Science, 333(6043), 776-778.

227 Carr, N. (2010). The shallows: What the Internet is doing to our brains. W. W. Norton & Company.

228 Uncapher, M. R., & Wagner, A. D. (2018). Minds and brains of media multitaskers: Current findings and future directions. Proceedings of the National Academy of Sciences, 115(40), 9889-9896.

229 Merchant, Z., Goetz, E. T., Cifuentes, L., Keeney-Kennicutt, W., & Davis, T. J. (2014). Effectiveness of virtual reality-based instruction on students' learning outcomes in K-12 and higher education: A meta-analysis. Computers & Education, 70, 29-40.

230 Radianti, J., Majchrzak, T. A., Fromm, J., & Wohlgenannt, I. (2020). A systematic review of immersive virtual reality applications for higher education: Design elements, lessons learned, and research agenda. Computers & Education, 147, 103778.

231 Means, B., Toyama, Y., Murphy, R., Bakia, M., & Jones, K. (2009). Evaluation of evidence-based practices in online learning: A meta-analysis and review of online learning studies. U.S. Department of Education.

232 Dellermann, D., Ebel, P., Söllner, M., & Leimeister, J. M. (2019). Hybrid intelligence. Business & Information Systems Engineering, 61(5), 637-643.

233 Jarrahi, M. H. (2018). Artificial intelligence and the future of work: Human-AI symbiosis in organizational decision making. Business Horizons, 61(4), 577-586.

234 Hancock, P. A., Billings, D. R., Schaefer, K. E., Chen, J. Y., De Visser, E. J., & Parasuraman, R. (2011). A meta-analysis of factors affecting trust in human-robot interaction. Human Factors, 53(5), 517-527.

235 Frey, C. B., & Osborne, M. A. (2017). The future of employment: How susceptible are jobs to computerisation? Technological Forecasting and Social Change, 114, 254-280.

236 The Future of Jobs Report 2020. (2020). World Economic Forum. https://www.weforum.org/publications/the-future-of-jobs-report-2020/

237 Brynjolfsson, E., & McAfee, A. (2014). The second machine age: Work, progress, and prosperity in a time of brilliant technologies. W W Norton & Co.

Chapter 8

238   Moore JW. What Is the Sense of Agency and Why Does it Matter?. Front Psychol. 2016;7:1272. Published 2016 Aug 29. doi:10.3389/fpsyg.2016.01272. Web. 11 Dec. 2019. https://www.ncbi. nlm.nih.gov/pmc/articles/PMC5002400/

239   Berman, R. Big Think. "A home library can have a powerful effect on children." Big Think. 13 Oct. 2018. Web. 11 Dec. 2019. https://bigthink.com/mind-brain/mind-brain-home-library-benefits

240   Sikora, J., Evans, M. D. R., & Kelley, J. (2019). Scholarly culture: How books in adolescence enhance adult literacy, numeracy and technology skills in 31 societies. Social Science Research, 77, 1–15. https://doi.org/10.1016/j.ssresearch.2018.10.003

241   Wharton Public Policy Initiative. "Bridging the Digital Divide." Wharton Public Policy Initiative. n.d. Web. 11 Dec. 2019. <https://publicpolicy.wharton.upenn.edu/live/news/2420-bridging-the-digital-divide/for-students/blog/news>

242   Johnson, S. D. and Aragon, S. R. "An Instructional Strategy Framework for Online Learning Environments," New Directions for Adult and Continuing Education 2003, no. 100, 31–43, https://doi.org/10.1002/ace.117. Wiley Online Library. 20 Nov. 2003. Web. 11 Dec. 2019. https://onlinelibrary.wiley.com/doi/10.1002/ace.117

243   Wikipedia Contributors. "Agency (sociology)." Wikipedia, the Free Encyclopedia. Wikimedia Foundation, 8 Dec. 2019. Web. 11 Dec. 2019. <https://en.wikipedia.org/wiki/Agency_(sociology)>

244   Bandura, A. "Self-Efficacy: The Exercise of Control – Chapter 2." Uky.edu. 15 Feb. 2013. Web. 11 Dec. 2019. https://www.uky.edu/~eushe2/Pajares/ebook2.html

245   Moore J. W. (2016). What Is the Sense of Agency and Why Does it Matter?. Frontiers in psychology, 7, 1272. doi:10.3389/fpsyg.2016.01272. Web. 11 Dec. 2019. <https://www.ncbi.nlm.nih.gov/pmc/articles/PMC5002400/>

246   Berberian B, Sarrazin JC, Le Blaye P, Haggard P. "Automation technology and sense of control: a window on human agency. – PubMed – NCBI." PLoS One. 2012;7(3):e34075. doi: 10.1371/journal.pone.0034075. Epub 2012 Mar 30. Web. 11 Dec. 2019. https://www.ncbi.nlm.nih.gov/pubmed/22479528

247   Artificial Intelligence and the Future of Teaching and Learning - Office of Educational Technology. (2023, May 24). Office of Educational Technology. https://tech.ed.gov/ai-future-of-teaching-and-learning/

248   State EdTech Trends - Leadership, Technology, Innovation, Learning | SETDA. (2024, September 11). Leadership, Technology, Innovation, Learning | SETDA. https://www.setda.org/priorities/state-trends/

249   Originally published: Shippee, M. (2023, October 25). The importance of agency in successful edtech adoptions. ESchool News. https://www.eschoolnews.com/innovative-teaching/2023/10/25/agency-successful-edtech-adoptions

250   Trophic Cascade | Definition, Importance, & Examples | Britannica. (2021). In Encyclopedia Britannica. https://www.britannica.com/science/trophic-cascade

251   Silliman, B. R. & Angelini, C. (2012) Trophic Cascades Across Diverse Plant Ecosystems. Nature Education Knowledge 3(10):44 https://www.nature.com/scitable/knowledge/library/trophic-cascades-across-diverse-plant-ecosystems-80060347/

252   Wolf Restoration – Yellowstone National Park (U.S. National Park Service). (2011). Nps.gov. https://www.nps.gov/yell/learn/nature/wolf-restoration.htm#:~:text=Much%20of%20the%20 wolves'%20prey,late%201800s%20and%20early%201900s.

Index

About the Author

Micah Shippee, PhD is the Director of Education Solutions at Samsung. Micah and his team design, develop, and deliver learning solutions to inspire and empower educators and learners. Micah strives to amplify the meaningful work being done in education by supporting the adoption of pragmatic innovation. Micah operates at the intersection of practice and research as a veteran consultant and professor specializing in planned change and innovation, learning theory, project management, and organizational behavior. His efforts focus on the adoption and deployment of new technological innovations in organizations. As an author, consultant, and keynote speaker, he focuses on the adoption of innovation through the development of cultures that embrace change. Micah is the author of *WanderlustEDU: An Educator's Guide to Innovation, Change, and Adventure* and co-author of *Reality Bytes: Innovative Learning Using Augmented and Virtual Reality.*